P9-CEH-751

Praise for *Oil and Honey*

"In this eloquent memoir, [McKibben] interweaves reportage on deluges, heatwaves, and melts with demonstrated solutions to 'malfunctioning modernity.' High-profile protest is only part of that, he argues. A revolution in local sustainability is also essential—and achievable, as the story of a Vermont beekeeper reveals." —*Nature*

"McKibben's book is . . . the chronicle of two sides of the environmental battle—the intensely personal and local, which centers on individual responsibility and humility, and the national and political, which hinges on mass movements, logistics, and existential solidarity . . . Whether we citizens as individual human beings can alter our fate is unknown. Political action, civil disobedience, divestment movements, recycling, local food movements, and public demonstration will all be required." —*The Wichita Eagle*

"Activist though he may be, McKibben remains a fine writer, evocative, articulate, clever, and humble in examining his mistakes. . . . Highly literate and expert musings on climate change, from the home to the global level."

—*Shelf Awareness*

"Tracking the emotional and intellectual journey that took McKibben from Vermont to picket lines in Washington, D.C., to town halls, universities, and arenas, the book is a call to action and an inspiring playbook for making change—both locally and globally—in the 21st century."

—*Publishers Weekly*

"Confiding and dramatic . . . In this moving, wryly amusing account set against the heated presidential debate of 2012, McKibben describes his extraordinary world travels and what it took to launch gutsy, creative, and effective protests, and shares invaluable information and such intriguing insights as what bees can teach us about reaching consensus. Galvanizing and inspiring." —*Booklist* (starred review)

"From the founder of the environmental organization 350.org, a chatty, warm memoir of his double life as globe-trotting activist and part-time novice beekeeper . . . A personal, enjoyably rancor-free account, filled with praise for his colleagues and some pokes at opponents but void of harangues."

—*Kirkus Reviews*

"Moving . . . McKibben's story will appeal to a broad range of readers, from those with general interest in environmental affairs and social movements to those committed to environmental protection and the power of moral witness in our society." —*Library Journal*

ALSO BY BILL McKIBBEN

Eaarth

The Bill McKibben Reader

Fight Global Warming Now

Deep Economy

Wandering Home

Enough

Long Distance

Hundred Dollar Holiday

Maybe One

The Comforting Whirlwind

Hope, Human and Wild

The Age of Missing Information

The End of Nature

OIL AND HONEY

THE EDUCATION OF
AN UNLIKELY ACTIVIST

BILL McKIBBEN

ST. MARTIN'S GRIFFIN   NEW YORK

Text stock contains 20% post-consumer waste recycled fiber

Printed in the United States of America. For information, address St. Martin's Press, 175 Fifth Avenue, New York, N.Y. 10010.

www.stmartins.com

Designed by Kelly S. Too

The Library of Congress has cataloged the Henry Holt edition as follows:

McKibben, Bill.
Oil and honey : the education of an unlikely activist / Bill McKibben.
p. cm.
ISBN 978-0-8050-9284-4 (hardcover)
ISBN 978-0-8050-9838-9 (e-book)
1. McKibben, Bill. 2. Environmentalism—United States. 3. Climatic changes—Environmental aspects. 4. Petroleum industry and trade—Environmental aspects. 5. Petroleum industry and trade—Political aspects—United States. 6. Environmentalists—United States—Biography. 7. Beekeepers—United States—Biography. I. Title.
GE197.M356 2013
363.70092—dc23
[B] 2013010995

ISBN 978-1-250-04871-4 (trade paperback)

St. Martin's Griffin books may be purchased for educational, business, or promotional use. For information on bulk purchases, please contact Macmillan Corporate and Premium Sales Department at 1-800-221-7945, extension 5442, or write specialmarkets@macmillan.com.

First published in hardcover by Times Books, an imprint of Henry Holt and Company

First St. Martin's Griffin Edition: August 2014

10 9 8 7 6 5 4 3 2 1

The farm is for Sophie, the fight is for Sophie,
and this book is for Sophie

CONTENTS

OIL AND HONEY

1

. . . . .

TWO LIVES

Here's a story of two lives lived in response to a crazy time—a time when the Arctic melted and the temperature soared, a time when the planet began to come apart, a time when bee populations suddenly dropped in half. Each story is extreme. They're not intended as suggestions for how others should live, and I hope the reader won't feel the need to choose, or reject, either one. Each story is mine, at least in part, for sometimes I think I've learned more in the past two years than in all the decades that came before. Some of that education came in the tumult and conflict of my own life, as I helped to build an active resistance to the fossil fuel industry. And some came in the beeyards of my home state, while I carefully watched a very different, very beautiful way of dealing with a malfunctioning modernity. These stories mesh together, I hope: awkwardly right now, but perhaps, with luck, more easily in the time to come.

• • •

I first met Kirk Webster in the fall of 2001. Newly ensconced at Middlebury College in Vermont, I'd offered to teach a course on local food production. There were two problems. One, I can't really grow anything—my heart is green, but not my thumb. Two, this was long before Michael Pollan or Barbara Kingsolver had taken up local agriculture, and there wasn't really much to read. We could choose among the remarkable essays of Wendell Berry, the seductive novels of Wendell Berry, and the tough poems of Wendell Berry. Looking through back issues of a magazine called *Small Farmer's Journal*, however, I came across an essay by a beekeeper named Kirk Webster. I'm not sure I noticed, the first time I read it, that he was a neighbor. I was just taken by his confident prose and his descriptions of his life among the honeybees.

"Surely the best kept secret in the U.S. today is the wonderful way of life that's possible with full-time farming on a small place," he began. "If more people understood the opportunities for faith, freedom, responsibility, health and education that good farming can provide, our rural areas might be repopulated and the self-destructive course of our society reversed. This timeless activity is so much more than just a way of making a living—it is in fact the Middle Path described in the Buddha's teachings and the object of St. Thomas's words: 'The kingdom of heaven surrounds you, but you see it not.'"

He was, it turned out, living in the next town over, and easy to track down via the small-farmer grapevine; he agreed to come to class and talk. I don't recall everything he said that day, but I do remember my first impression: he was bearded, shy, and a little ill at ease, but we all took to him instantly. Even the students who had no intention of becoming farmers—the ones bound for finance or medicine or the other high-powered careers you leave for from a place like Middlebury—were shaken a little by his quiet resolution, and by his story.

He'd grown up in suburban New Jersey (like many of them), in a family he described in his essay as "largely dysfunctional and aimless" (so, not unlike a lot of them). "I always liked to read, and I didn't have trouble getting good grades, so everyone assumed I would be able to get scholarships and somehow continue as far as possible with 'education.'" By the age of fifteen, though, "it was clear that I was soon going to seek elsewhere for something to do in my life." Nature and the outdoors world had become an "irresistible magnet," and so in order that he earn some kind of diploma his parents sent him to the Mountain School in farm country Vermont, a rural outpost that grew its own food and cut its own firewood, and where he was all but adopted by one of the families whose parents taught at the school. Bill and Martha Treichler, and their boys and girls, taught him how to garden and to build and to do the hundreds of other jobs of rural self-sufficiency; he suddenly had a model that made sense—a joyful and tight farm family who were living outside the normal economy.

"One evening, just before dinner in the noisy school dining hall," he wrote, "Bill told me that the year their fifth child was born, the family's gross income was $600. I almost dropped the pitcher of milk I was holding. The sights and sounds in the room started to spin, and I felt like someone had just hit me right between the eyes with a stick of cordwood. Here were the most capable, healthiest, and best educated people I had ever met, who with five young children at home, had chosen a way of life with only $600 of cash income (perhaps equivalent to $2,400 today). They certainly could have pursued any number of jobs or careers to make a normal income, but chose instead to be together as a family and pick and choose carefully which aspects of the larger society they would get involved with. Farming and healthy self-sufficient living in a debt-free situation allowed them to do this. In that moment in the dining hall, all of my developing notions of making a living, security, jobs and careers were shattered, and I knew I would have to start again in learning what these things really mean."

That moment ramified. When he was wracked up in a toboggan accident that winter, someone gave him a book on beekeeping, and it captured his imagination; home on vacation in New Jersey he found an octogenarian Ukrainian immigrant who needed help with his hives. That man told him about another—Charlie Mraz, in Vermont's Champlain Valley, and when Kirk returned to school he hitchhiked across the state to ask the veteran apiarist for a job. He worked there for two years after high school, eating meals with the family but sleeping in the honey house. And then, still a very young man, he struck off across the country, working on a variety of farms and doing carpentry to pay his bills. Everywhere he went he built up small apiaries, honing his skills, and in the fall of 1985 he returned to the Champlain Valley and began his life's work, raising bees and selling colonies, queens, and honey. Slowly, patiently, and in the face of growing problems with mites that were decimating many apiaries, he built his business into a going concern, pioneering a number of new techniques and becoming one of the very few beekeepers in the country who made a living without using chemicals in his hives. It was a decent living, too—when he came to my class that day, he bought his books with him, and showed us that Champlain Valley Bees and Queens, Inc., was grossing $50,000 a year, of which about half netted out. "After living, and enjoying life, for so long with so little money, this frankly seems like an enormous fortune to me," he said. "In terms of the American greedy lifestyle, it's still not very much money. But I consider it to be a more than ample reward for the independence, the wonderful way of life, and the chance to live apart from a predatory society that beekeeping and farming provide."

He was, in other words, leading a somewhat Amish life, with the obvious exception that he wasn't surrounded by an Amish community where everyone else was living likewise. There are other small farmers in the valley, and they were his friends; nonetheless, he was, perhaps, a little lonely—more on that later. But the deeper problem went like this: he thought his farming wouldn't truly matter until he could pass on what he'd learned. "If there are young people anymore, interested in beekeeping, I'd like to have a few of them come here to learn the trade," he wrote. "This is still in the planning stage, but it should be possible to expand the apiary enough to support one or two apprentices, then spin off the excess bees as the young folks return home to start propagating bees and producing honey on their own. If even one or two full-time apiaries resulted from this process, I'd be able to at least approach my own definition of successful farming."

As the decade wore on, I'd see Kirk now and again—have him over for dinner or meet him for a cross-country ski. And so I knew he was shepherding his apiary through the most difficult decade in beekeeping history, surviving everything from the colony collapse disorder that killed so many beehives to the flood of cheap (and adulterated) Chinese honey that threatened to wreck the market. He'd continued to follow his unorthodox route. Instead of trucking his bees to California, like most apiarists, to cash in on the almond pollination season, he kept them close to home all year round, and worked diligently to rid his apiary of all trace of chemicals. And it had worked—but not well enough for him to take on the apprentices he'd wanted. He had no farm of his own, so he lived in a rented home on a small patch of land and had his shop nearby; his colonies were, as with most apiaries, spread out at a dozen locations around the valley. It all worked, but there was no room for young people to come, stay, and learn. And there was no land to make the apiary the hub of something even sweeter, a small farm with crops and animals. Had he lived some other place, he could have done it, but the cost of land in Vermont is unnaturally high—New York and Boston are within driving distance, and so prices get set less by what a farmer can earn than by what a stockbroker can afford.

It became clear to me that the moment was passing—Kirk is strong and healthy, but he's got another decade at his peak, I'd guess. If he was going to pass on what he knew, the time was ripe. And I, too, felt a strong urge to have a more-than-theoretical connection to the landscape and the emerging local economy that I was writing so much about. So I made him a proposal: What if I buy you a piece of land and grant you free lifetime tenure on it? In return, you build the farm buildings and get the land working, and pay the insurance and taxes. By any global standard, I'm a rich man. But I'm not in the class of people who buy farms willy-nilly. Still, I've always wanted something tangible to leave my daughter; since Kirk and I are about the same age, she should be the ultimate beneficiary, inheriting the operation when Kirk died. Given what

I knew about climate change, the gift of productive land seemed like the best thing I could hope to pass on to her, an insurance policy worth more than money in some account. In the meantime, Kirk could fulfill his farming destiny.

Kirk agreed, and I went looking for the money—as it turns out, the check for this book covered the down payment. And together we started the search for land, wandering one property after another. There was no shortage of possibilities—every month a few more dairy farms disappear, done in by the low price of commodity milk and the impossibility of competing with the giant ten-thousand-head megadairies of the West. We looked at many, but they were hard worn, their outbuildings crumbling after a few decades of cash-strapped deferred maintenance. We eventually checked in with the Vermont Land Trust, which has been conserving farmland around the state for decades. (It works like this: a farmer decides that instead of selling off his land in lots for vacation homes, he'll sell the development rights to VLT; he can keep farming, and the land will stay intact.) VLT connected us with a farmer who wanted to unload—after selling his development rights he'd gotten sick of the entire farming business altogether and moved on to California, and now his seventy-acre parcel outside the town of New Haven was just sitting there. There was a driveway and one double-wide trailer. The land was pretty near the geographic center of Kirk's various beeyards around the county, and when we tested the well the water flowed pretty well. With the great help of our lawyer friend Dick Foote we managed finally to settle the deal. The farm wasn't especially picturesque—the neighbor directly to the west ran a noisy excavating business, and the fields were rimmed with scrubby sumac. But some of the soil was rich loam, not the standard Champlain Valley clay. And the woodlot was plenty large enough to keep Kirk in firewood forever. We both knew it was the place.

The double-wide would serve for the someday apprentices; the first order of business, in that spring of 2011, was to get a barn built, and then, if his money held out, a small farmhouse, where Kirk was pretty sure he'd spend the rest of his life. This new operation would not change the world, both of us knew that. But it would, you know, change the world. The sum total of a million of these kind of small shifts would be a different civilization, one you could just begin to sense emerging as farmer's markets spread across the nation. The U.S. Department of Agriculture had just announced a seismic demographic shift: For the first time in 150 years the number of farms in America was no longer falling. In fact, over the past half decade, it had begun inexorably to rise. All the growth was coming at the small end of the business, with people growing food for their neighbors. Vermont was a case in point: dairies continued to disappear, but we suddenly had neighbors growing wheat and barley—the kind of crops we hadn't seen for a century in this state. The number of farmers in the United States was still small—just 1 percent, or half the proportion of the population behind prison bars. But something had definitely begun to turn. Given enough time . . .

Time, of course, was the trouble. Offered a century's grace, I have no doubt we could subside into a workable, even beautiful, civilization. But 2011, when Kirk and I bought the farm, was shaping up to be one of the warmest years on record. As that summer wore on, we saw record heat in the Southwest and a drought so deep it killed five hundred million trees in Texas. Meanwhile, there was record rainfall across the Mississippi Basin, and the river swelled so fast that the Army Corps of Engineers was blowing up levees and flooding farmland to try to save cities from inundation.

Those were the facts of my life, those and a million other such stories and statistics. For twenty-five years—almost my entire adulthood—I'd been working on what we first called the greenhouse effect, and then global warming, and then climate change. Back in 1989, when Kirk was building his first apiaries, I was writing my first book, which was also the first book on the topic for nonscientists. *The End of Nature* was a best seller, translated into a couple of dozen languages, and my initial theory (I was still in my twenties) was that people would read the book—and then change.

That's not quite how it happened, so I kept on writing, one book after another, about some aspect of this great crisis. I wrote articles, too, for just about every magazine you could name, and op-eds, and when blog posts became a thing I wrote those. I assumed, like most people, that reason would eventually prevail—that given the loud alarm sounded by scientists, governments would take care of the problem. And for a while that seemed, fitfully, to be happening. I was in Kyoto in 1998 when the world's nations signed the first accord to staunch the flow of carbon dioxide, and I remember thinking that we'd turned a corner. It was going to be close, I thought, but we were headed in the right direction.

That's not quite how it happened, either. As it turned out, the United States never ratified the Kyoto accord, and soon China was building a coal plant a week. Carbon emissions kept soaring, and donations from the fossil fuel industry managed to turn one of our two political parties into climate deniers and the other party into cowards. Power, not reason, was ascendant, and writing yet another story about the latest scientific findings seemed less and less useful. By 2009, a decade after Kyoto, the U.S. Senate—then with sixty Democrats—was so scared of Big Oil that it wouldn't even take a vote on the most modest, tepid climate legislation imaginable. And six months later the world convened in Copenhagen for a failed climate summit that killed any hope of global progress.

Sometime in the course of the past decade I figured out that I needed to do more than write—if this fight was about power, then we who wanted change had to assemble some. Environmentalists clearly weren't going to outspend the fossil fuel industry, so we'd need to find other currencies: the currencies of movement. Instead of money, passion; instead of money, numbers; instead of money, creativity.

At first—this was 2006—I had no clue at all. I called a few Vermont writer friends of mine, and asked if they'd come to our main city, Burlington, and sit in on the steps of the federal building. We'd be arrested, there'd be a small story in the paper, we'd have done *something*. They agreed—but one of them called the police and asked what would happen to us. "Nothing," was the reply. "Sit there as long as you want." So instead I asked people to walk across Vermont—we left from Robert Frost's old summer writing cabin, which is near my house, and walked for five days, sleeping in farm fields along the way. By the time we got to Burlington there were a thousand people marching, which in Vermont is a lot—enough, as it turned out, to get all our candidates for federal office (even the Republicans) to sign a pledge that they'd work in Congress to cut carbon emissions dramatically.

The next day, though, a newspaper account called that protest the largest demonstration against climate change that had yet taken place in the United States, and suddenly I understood better why we were losing. We had the superstructure of a movement: scientists, economists, policy experts, Al Gore. In fact, all we were lacking for a real movement was the movement part, the surge of people that produces respect and maybe even a little fear in leaders. Activists on the front lines were doing superb work fighting individual power plants and coal mines, but they weren't getting the support they needed—it wasn't adding up fast enough. So we set out to build one.

When I say "we," I mean me and a small team of undergraduates at Middlebury College, where I teach. We'd met one another in those long days of walking across Vermont, and I'd been deeply impressed by their budding talents and their good cheer. So that winter we launched a campaign called Step It Up, and in the course of three months created a springtime day of action that coordinated 1,400 protests across all fifty states. (The one in North Dakota was small.) We were successful in part because of beginner's luck and in part because my young colleagues knew more about the Internet than the rest of the environmental movement put together. Mostly, though, we were pushing on an open door—there were plenty of people who were deeply concerned about global warming but felt powerless in its face. When we finally offered them the chance to unite their voices, they took it eagerly. Both Barack Obama and Hillary Clinton, then running for president, took note of the rallies, and a few days later changed their platforms to reflect our goal: an 80 percent cut in carbon dioxide emissions by 2050. We were feeling . . . smug.

But a few weeks later, in the summer of 2007, the Arctic

began to melt, breaking all previous records. Clearly climate change was coming faster than even the most pessimistic scientists had thought, and 2050 was no longer all that relevant. We'd need to work faster, on a larger scale. NASA's James Hansen, the planet's premier climate scientist, provided us with a number: in January 2008, his team published a paper showing that if the concentration of carbon dioxide in the atmosphere rose above 350 parts per million, we couldn't have a planet "similar to the one on which civilization developed and to which life on earth is adapted." (Five years later we're closing in on 400 parts per million—that's why the Arctic is melting.) We took 350.org for our name, reasoning that we wanted to work all over the world (they don't call it global warming for nothing) and that Arabic numerals crossed linguistic boundaries. And then we took a leap of faith that in retrospect seems ludicrous—since there were seven continents, each of those seven young people working with me took a chunk of one and we set to work: Kelly Blynn on South America, Jeremy Osborn in Europe, Phil Aroneanu in Africa, Will Bates on the Indian subcontinent, Jamie Henn in the rest of Asia, May Boeve at home in North America, and Jon Warnow on the antipodes (he also got the Internet). Our success the year before meant that a couple of foundations (the Rockefellers, the Schumanns) were willing to help fund our work, and so the rest of the team was getting paid small salaries, and they had money to travel. But how do you just land in, say, Vietnam or Peru or Kazakhstan and start "organizing"? We found out.

The seven kids did endless work—literally endless, since going global meant there was always someone awake somewhere to e-mail. Mostly we found people like ourselves—there aren't "environmentalists" everywhere, but there's always someone worried about public health or hunger or war and peace. (Worried, that is, about all the hopes that will be wrecked if the planet starts to fail.) Though most of them were poor, and hence living lives a world apart from that of New England college students, they were natural allies, quick to understand both the science and the politics. And so by the fall of 2009, we were ready to hold our first global day of action. It was beyond exciting watching the pictures pour in—there were 5,200 rallies in 181 countries, what CNN called "the most widespread day of action in the planet's history." We followed it up with three more big global extravaganzas—thousands of demonstrations everywhere, save North Korea. There are forty thousand images in the Flickr account—I can show you pictures from Mongolia and Mumbai and Mozambique, from Montreal and Mombasa and Mauritania. From almost everywhere. We did our part to educate the world about what was coming at it.

But if you've built a movement, you've eventually got to put it to work. And now "eventually" had come. Education needed to yield to action.

So while Kirk was starting to build his barn in that early summer of 2011, I was stepping off a small cliff into the next phase of my life. To this point I'd been able to pretend that I was mostly a writer who happened to be helping with some activism—that our global climate education project was a natural extension of the work I'd spent my life doing. But now I was getting ready to do something different: to pick a tough, visible fight with the strongest possible adversaries on the biggest political stage in the world. Global warming was accelerating—2010 had just set the new record for the hottest year ever recorded. It was time to pick up the pace and move from engagement to resistance.

And so, at least for the two years described here, I've made the transition, however reluctantly, from author-activist to activist. Except for a few blessed interludes in the beeyards, I've spent my time on the computer and the airplane and the phone, giving speeches and leading marches. I've willed myself to be someone other than who I had been. The strain has told; I've changed, and not always for the best. This is the story of that education.

I miss, sometimes desperately, the other me: the one who knew lots about reason and beauty and very little about the way power works; the one with time to think. But time, as I say, is what we're lacking.

As it happened, I'd spent the spring of 2011 teaching a course at Middlebury. "Social Movements, Theory and Practice," it was called—but since these were the opening months of the Arab Spring it was mostly practice. We watched YouTube videos of young Egyptians organizing epic marches, and brave Libyans standing up to their tyrant Muammar Gadhafi. And we read Taylor Branch's classic three-volume biography of Martin Luther King Jr. and the civil rights years, which might as well be a handbook for organizers—it's so full of behind-the-scenes details that you could see exactly how Dr. King had dealt with every problem we'd face, from stubborn presidents to (far harder) stubborn colleagues from the large civil rights organizations.

By the time I was done with the semester, I'd decided that 350.org should organize the first major civil disobedience action for the climate movement. I sensed, from the speeches I was giving and the e-mail that flowed in hourly, that people were ready for a deeper challenge—it was time to stop changing lightbulbs and start changing systems. If we were going to shake things up, we'd need to use the power King had tapped: the power of direct action and unearned suffering. We'd need to go to jail.

And at precisely that moment, an issue materialized out of thin, if dirty, air. In the spring of 2011 Jim Hansen published a small paper pointing out that "peak oil" was not, in fact, happening quite as expected—that though we were indeed running out of easy-to-tap sweet crude, the newly emerging category of "unconventional oil," and in particular the tar sands of Canada, contained huge amounts of carbon. Those Albertan tar sands, he wrote, were so gigantic that if we burned them in addition to everything else we were burning, it would be "game over for the climate."

His calculations put a sudden spotlight on a previously little-known pipeline proposal called Keystone XL that was designed to carry almost a million barrels a day of that tar sands oil south from Canada to the Gulf of Mexico. Native leaders in Canada had been fighting tar sands mining for years, because it had wrecked their lands—only 3 percent of the oil had been pumped out, but already the world's biggest bulldozers and dump trucks had moved more earth than was moved building the planet's ten biggest dams, the Great Wall of China, and the Suez Canal combined. And some ranchers in the United States had begun to rally along the planned route of the pipeline itself, particularly in Nebraska, where it was destined to run straight across the iconic Sandhills and atop the Ogallala Aquifer that irrigates the Great Plains. But these protests hadn't gained enough traction to stop the plan. Keystone XL awaited only a presidential permit.

That was the part that interested me. An old law, mainly used for things such as building a bridge between New

Brunswick and Maine, required presidents to declare that any infrastructure crossing our country's border was "in the national interest." Congress didn't need to act, which was good since I knew there was no possible way to even think about convincing the Republican-controlled House of Representatives to block the pipeline. But this decision would be made by Barack Obama, and Barack Obama was fifteen months away from an election. Maybe we had an opening to apply some pressure—an opening to see if we'd nurtured a climate movement strong enough to make a difference.

And so I called the native leaders, who'd been fighting the longest, and asked if it was okay if we joined in. They graciously refrained from pointing out we were late to the game, and promised to collaborate (a promise they would keep in spectacular fashion in the year ahead). And then I called the small but hardy band of environmental campaigners in Nebraska and in Washington, D.C., who had been trying to block the pipeline. If we demanded more dramatic action, I asked, would it somehow damage their efforts? "We're losing," they said. "We have no deal for you to damage. Going to jail can't hurt."

Our small crew at 350.org—still run by those seven young people, now fully grown and highly able organizers—talked it through. We knew it was a gamble, but when you're behind, you take risks. (The slow, easy, sensible trajectories for dealing with climate change were in the past now; sometimes I had to restrain myself from saying to some "moderate" politician, "If only you'd listened to me a quarter century ago. . . .") And when you're losing you take personal risks, too; I sensed I was stepping over a line. With no idea how it would all come out, I sat down and wrote a letter, which I circulated to a few of my friends to cosign. It went out into the far reaches of the Web in June 2011, and it was as blunt as I could make it.

> Dear Friends,
>
> This will be a slightly longer letter than common for the Internet age—it's serious stuff.
>
> The short version is we want you to consider doing something hard: coming to Washington in the hottest and stickiest weeks of the summer and engaging in civil disobedience that will quite possibly get you arrested.
>
> The full version goes like this:
>
> As you know, the planet is steadily warming: 2010 was the warmest year on record, and we've seen the resulting chaos in almost every corner of the earth.
>
> And as you also know, our democracy is increasingly controlled by special interests interested only in their short-term profit.
>
> These two trends collide this summer in Washington, where the State Department and the White House have to decide whether to grant a certificate of "national interest" to some of the biggest fossil fuel players on Earth. These corporations want to build the so-called Keystone XL pipeline from Canada's tar sands to Texas refineries.
>
> To call this project a horror is serious understatement. The tar sands have wrecked huge parts of Alberta, disrupting ways of life in indigenous communities—First Nations communities in Canada and tribes along the pipeline route in the U.S. have demanded the destruction cease. The pipeline crosses crucial areas like the Ogallala Aquifer where a spill would be

disastrous—and though the pipeline companies insist they are using "state of the art" technologies that should leak only once every seven years, the precursor pipeline and its pumping stations have leaked a dozen times in the past year. These local impacts alone would be cause enough to block such a plan. But the Keystone pipeline would also be a fifteen hundred mile fuse to the biggest carbon bomb on the continent, a way to make it easier and faster to trigger the final overheating of our planet, the one place to which we are all indigenous.

As the climatologist Jim Hansen (one of the signatories to this letter) explained, if we have any chance of getting back to a stable climate "the principal requirement is that coal emissions must be phased out by 2030 and unconventional fossil fuels, such as tar sands, must be left in the ground." In other words, he added, "if the tar sands are thrown into the mix it is essentially game over." The Keystone pipeline is an essential part of the game. "Unless we get increased market access, like with Keystone XL, we're going to be stuck," Ralph Glass, an economist and vice president at AJM Petroleum Consultants in Calgary, told a Canadian newspaper last week.

Given all that, you'd suspect that there's no way the Obama administration would ever permit this pipeline. But in the last few months the administration has signed pieces of paper opening much of Alaska to oil drilling, and permitting coal mining on federal land in Wyoming that will produce as much CO2 as three hundred power plants operating at full bore.

And Secretary of State Clinton has already said she's "inclined" to recommend the pipeline go forward. Partly it's because of the political commotion over high gas prices, though more tar sands oil would do nothing to change that picture. But it's also because of intense pressure from industry.

So we're pretty sure that without serious pressure the Keystone pipeline will get its permit from Washington. A wonderful coalition of environmental groups has built a strong campaign across the continent—from Cree and Déné indigenous leaders to Nebraska farmers, they've spoken out strongly against the destruction of their land. We need to join them, and to say even if our own homes won't be crossed by this pipeline, our joint home—the earth—will be wrecked by the carbon that pours down it.

And we need to say something else, too: it's time to stop letting corporate power make the most important decisions our planet faces. We don't have the money to compete with those corporations, but we do have our bodies, and beginning in mid-August many of us will use them. We will, each day, march on the White House, risking arrest with our trespass. We will do it in dignified fashion, demonstrating that in this case we are the conservatives, and that our foes—who would change the composition of the atmosphere—are dangerous radicals. Come dressed as if for a business meeting—this is, in fact, serious business.

And another sartorial tip—if you wore an Obama button during the 2008 campaign, why not wear it again? We very much still want to believe in the promise of that young senator who told us that with his election the "rise of the oceans would begin to slow and the planet start to heal." We don't understand what combination of bureaucratic obstinacy and insider dealing has derailed those efforts, but we remember his request that his supporters continue on after the election to pressure his government for change. We'll do what we can.

One more thing: we don't just want college kids to be the participants in this fight. They've led the way so far on climate change—ten thousand came to D.C. for the Power Shift gathering earlier this spring. They've marched this month in West Virginia to protest mountaintop removal; a young man named Tim DeChristopher faces sentencing this summer in Utah for his creative protest.

Now it's time for people who've spent their lives pouring carbon into the atmosphere to step up, too, just as many of us did in earlier battles for civil rights or for peace. Most of us signing this letter are veterans of this work, and we think it's past time for elders to behave like elders. One thing we don't want is a smash up: if you can't control your passions, this action is not for you.

This won't be a one-shot day of action. We plan for it to continue for several weeks, till the administration understands we won't go away. Not all of us can actually get arrested—half the signatories to this letter live in Canada, and might well find our entry into the U.S. barred. But we will be making plans for sympathy demonstrations outside Canadian consulates in the U.S., and U.S. consulates in Canada—the decision makers need to know they're being watched.

Twenty years of patiently explaining the climate crisis to our leaders hasn't worked. Maybe moral witness will help. You have to start somewhere, and we choose here and now.

We know we're asking a lot. You should think long and hard on it, and pray if you're the praying type. But to us, it's as much privilege as burden to get to join this fight in the most serious possible way. We hope you'll join us.

Maude Barlow—Chair, Council of Canadians
Wendell Berry—Author and Farmer
Danny Glover—Actor
Tom Goldtooth—Director, Indigenous Environmental Network
James Hansen—Climate Scientist

Wes Jackson—Agronomist, President of the Land Institute
Naomi Klein—Author and Journalist
Bill McKibben—Writer and Environmentalist
George Poitras—Mikisew Cree First Nation
Gus Speth—Environmental Lawyer and Activist
David Suzuki—Scientist, Environmentalist, and Broadcaster
Joseph B. Uehlein—Labor Organizer and Environmentalist

It's the kind of letter where you sit there with your hand above the send button and just kind of wonder how much your life is going to change. As it turned out, a lot.

One reason I find it hard to ask people to come to D.C. and get arrested: part of me thinks, strongly, we should stay home. Or at least that I should.

Yes, because it takes energy to travel. Yes, it has, in fact, occurred to me that there's something remarkably ironic about my flying around the world to build a climate movement. I do it because I think the math works: if we can stop Keystone, that's nine hundred thousand barrels a day for the fifty-year life of the pipeline. But it always nags at me, that surge of power at the top of the runway as the jet engines guzzle fuel to get us aloft. I tell myself that we fight this fight in the world we live in, not the one we hope to build.

But we *do* need to build that world, and that's even more why we should stay home; it's why Kirk's project attracts me. It's clear to me that we can't have precisely the same economy that we've grown up with, not the globe-spanning anything-at-any-time consumerism, not the starter-castles-for-entry-level-monarchs housing stock, not the every-man-a-Denali/Tahoe/Escalade driveway. We're going to have to change our patterns, our laws, our economies, our expectations. My last few books have focused on the possibilities for local food, local energy, local currency—and the appeal for me is not just or even mainly intellectual.

I found the mountains surrounding Lake Champlain as a fairly young man. I'd moved to the Adirondacks at the age of twenty-six, falling in love with the sheer wildness of the place, a bigger tract of protected land than Glacier, the Grand Canyon, Yellowstone, and Yosemite combined. A child of the suburbs, I was knocked over by the contact with hot, cold, wet; it was no different than any other incandescent young love, except that it has burned on for years. Hemlock bowing across the stream, red pine needles baking in the August sun high on the ridge, coyotes yipping in the night, sky so black the Milky Way stretches to each horizon; all of it was a revelation. (In fact, the dominant emotion of *The End of Nature* was not fear but sadness—a lament for the wildness that climate change threatened to leach away.) Just to say the names calms me down: Ampersand Mountain, Thirteenth Lake, Raquette River.

After fifteen years my wife and my daughter and I moved fifty miles across the lake to Vermont, sacrificing a little of the wildness for the strong sense of community that defines the Green Mountain State: the town meeting, the farmer's market, the microbrewery. These are places that might be made to work: the Adirondacks is the best example of a wilderness with people living in and among it; Vermont the best example of an earlier American state of mind, before the hyper-individualism of the TV age completely took over.

There's so much to be done here at home; you can sense the new world coming into embryonic form, with its own sources of everything from seeds to capital. And for me, even more, it's the landscape that fits with jigsaw precision into the hole in my heart. I'm happy when I'm home, when I can see the sun shining through the winter-bare ridge at dusk, when I can swat the blackflies come June. My thirties were essentially an extended early retirement; I spent those years—the 1990s—writing and wandering, and watching my daughter grow. Her first word was "birch," which pleased me more than I can say; by the time she was fifteen she'd climbed all forty-six of the high peaks in her native Adirondacks, which made me at least as proud as her college admissions letter did a few years later.

Which is why it's so odd that I've spent more nights away than home these past years. I've been to every continent since 2008, and once I hit four of them in six days. At 350.org we've organized, in the words of *Outside* magazine, "more rallies than Lenin and Gandhi and Martin Luther King combined." It's been the most satisfying work of my life, endlessly difficult and endlessly interesting. But asleep in some Days Inn or Courtyard by Marriott, I dreamed of the Champlain Valley, with the Adirondacks towering to the west and its growing web of organic dairies and community-supported agriculture (CSA) farms; I woke up to eat at the breakfast bar (non-Vermont non-maple syrup) and do rhetorical battle with retrograde congressmen. But I did that battle in the name of my place, remembering what it felt like. I can try to imagine "unborn generations" and the "suffering poor" and the other huge reasons to fight climate change, but I never have the slightest trouble conjuring up the tang of the first frosty morning in the Adirondack fall, the evening breeze that stirs as the sun drops below the ridge.

And, of course, if I knew my place, Kirk *really* knew

it—felt its every change not only with his own senses but with the extended vision of the many million bees in his charge. Through them he knew each new development in the wider world; they were scouts, and he could read their dispatches with ease. I know no one more connected, which is why it has been a privilege just to follow him around.

So when I say activism didn't come naturally to me, it's not simply because I'm a writer; it's because the need to stay close to home was very nearly biological. If I missed a week wandering the woods, it meant not seeing those flowers that year—the trillium would have to wait till next spring. But I'd turned fifty, and the "next springs" were now fewer than the springs I'd known. At night, on the road, distracted by worry, I'd say those names: Camel's Hump, Breadloaf Mountain, Otter Creek. I'm a mediocre meditator, but the one mantra that could lull me to sleep in some lonely Hilton was the list of Lake Champlain's many tributaries, north to south along the Vermont shore, then back down on the New York side. It hurt, physically, to leave; flying back into Burlington airport, winging past Whiteface and Giant Mountain, wheeling over Missisquoi Bay, calmed me down like nothing else.

That the two sides of my life were so at odds bothered me no end, far more than the jet fuel my travels burned. I couldn't quite make them connect.

2

STORMS

We weren't really planning to actually go to jail.

Our advance team had been on the ground in Washington for three weeks. It turns out that in a market society there are people equipped to fill every need, including organizing civil disobedience. The crew we'd found, and who would soon become close colleagues, was headed by Matt Leonard. With his shaved head and earring, he bore a striking resemblance to Mr. Clean. But he was more like Mr. Calm; in what became a rapidly mounting storm he never lost his cool. His posse included Rae Breaux, Linda Capato, Duncan Meisel, and Josh Kahn Russell, each the veteran of many such actions.

But because of that history, they were pretty sure only a small number of people would turn up—it had been at least thirty years since people had been hauled away in the thousands. (That had occurred during protests at the Rocky Flats nuclear test site in the Western desert.) Matt kept saying that we'd be fine with five or ten arrestees a day over our two-week protest; even as the number of people signing up kept mounting, he cautioned that many would melt away. The D.C. police must have felt the same way, because it was next to impossible to get their attention—our team was bounced from one sergeant to another, and none seemed to take the whole thing very seriously. I began to worry they'd just let us sit there, that we wouldn't get arrested at all.

We'd told each daily wave of potential arrestees to gather the night before at a Washington church for training. So it wasn't until five p.m. on Friday, August 19, 2011, when we convened the first of these sessions at St. Stephen and the Incarnation Episcopal Church in the Columbia Heights neighborhood, that we got to find out if anyone would really show. People started streaming in early; soon there were more than eighty people planning to risk arrest the next day, which was more than we'd anticipated—many more. We practiced for an hour, walking in columns down the center aisle and fanning out in front of the altar as if it were the White House lawn. A procession of lawyers from the National Lawyers Guild answered questions ("What if I have a green card?" "Will they take my medicine?") and assured us that the routine for arrests was well established. We'd sit down on the sidewalk directly in front of the White House in a fifty-yard zone called the "postcard window" reserved for people taking snapshots. The police would handcuff us, load us in paddy wagons, drive us to the police station, process us, fine us a hundred dollars apiece, and release us that afternoon: we even handed out slips of paper with subway directions from the police station back to the airports and train stations, because most people were planning to travel home that evening.

But it turned out that the police were not as pleased by our turnout as we were. When we arrived the next morning at Lafayette Square, across the street from the White House, there were a scattering of police cars—but as the crowd swelled, more and more cops kept arriving, too. And though the rest of us didn't know it at the time, an angry lieutenant was giving Matt the first indication that the arrests might be anything but routine.

I made a short speech through a bullhorn, and then, as we'd practiced the night before, we walked toward the White House, spreading out into a long line three deep, with our banner at the center. We sat on the sidewalk, the president's front porch behind us, and then we waited. People tried a few chants, but they didn't fit the mood, which seemed solemn, but also joyful. To me at least it felt like I was finally taking action on a scale that began to match the scale of the problem—that if the planet was at stake, handcuffs made more sense than lightbulbs. I was grinning, I think.

Three times a police officer read out an order to leave: "Attention. This is Lieutenant Phelps of the U.S. Park Police. Because of violations of park laws and regulations applicable to this area, your permit to demonstrate on the White House sidewalk has been revoked by the ranking supervisory U.S. Park Police official in charge. Due to these violations the sidewalk is closed. All persons remaining on the closed portion of the White House sidewalk will be arrested. This is your third warning." With that, they closed metal barricades around us, shooed away the onlookers and camera crews, and began arresting us, one by one. It took a long time (which, as it turned out, would be the basic operating principle for the next few days). Beginning with the women, officers in full body armor hauled us one by one to our feet, cuffed our hands behind our backs, and then led us to a small tent, where they photographed us and gave us each a number. From there we were escorted to the back of a paddy wagon, which was stifling hot and claustrophobic once it filled with the requisite ten bodies. And from there we took a ten-minute drive to the U.S. Park Police station in Anacostia.

We sat on the ground outside the station for an hour or two, hands still cuffed behind our backs—after a while, it's painful. And then, one by one, we were led inside, where an officer emptied our pockets and wrote out in laborious longhand a receipt for each of us. (The Park Police seem not to have been informed about the advent of the digital age—I did see a couple of IBM Selectrics on a desk, but they were unplugged. It was pure Bic and carbon paper for us.) That's when my wedding ring went, along with my ID, my hundred dollars, my belt, my shoelaces, and my necktie. But I really only cared about the wedding ring—it's amazing how much you suddenly miss something that normally you don't even know you're wearing.

"Why are you taking it?" I asked.

"Because where we're sending you they'll cut your finger off to get it," one cop explained.

That was a pretty good clue, but we still didn't fully understand we were headed for jail. The lawyers had been so sure it would be just a matter of hours before we'd be released back out into the sweltering Washington afternoon—everyone still had their slips of paper with the directions to the airport. But as the afternoon dragged on, the police broke us into groups of ten or fifteen and ushered us into small holding cells, small enough that only one or two could sit on the floor at a time while the rest stood around them. (There was a stainless steel toilet in each cell, too, and it took a while before we overcame our reluctance to use it while surrounded by a dozen others.) We stood for hours, and it gradually began to dawn on us that we were not, in fact, going anywhere soon. They let us out one at a time to make our single phone calls, just like in the movies—I called our support team working out of a borrowed office in D.C. and learned the bad news: we were almost certainly going to be held overnight, and probably the next night, too. The police had told Matt they didn't want to deal with two weeks of demonstrations of this size, so they were upping the price to try to deter those planning to come.

This scared me—not the jail part (well, a little), but the deterrence part. What if it worked? I mean, we'd been telling people that the most likely outcome was a few hours in a police station, not a few days in jail. Would the protest just fizzle now? I asked our communications coordinator, Jamie Henn, if he could spread around a one-sentence message: "We don't need sympathy; we need company." And that was the last message I'd manage to get out for the duration.

I'd spent all summer plotting to get us arrested, so there was no point in complaining. About ten o'clock that Saturday night they loaded us back into the paddy wagon and took us to D.C.'s Central Cell Block. They put me in a holding room and shackled my ankle to a bolt in the floor, then fingerprinted me and led me to a cell: a small steel cage with two steel bunks and a steel toilet/sink. There are, one hears, "country club prisons," but not, I think, country club jails—this one lacked mattresses, pillows, even sheets. Just stainless steel. My cell mate—Curt, a real estate salesman and blogger from Louisville, Kentucky, who had driven east for the protest—was on the upper bunk, a much tougher spot since the light shone all night in his face. We chatted, and we talked through the bars to find out who was nearby: Gus Speth, one of the great heroes of the environmental movement, was in the next cell; Jim Antal, conference minister for the United Church of Christ in Massachusetts, was a few doors down; next to him was Chris Shaw, one of my oldest hiking and paddling and writing friends from the woods of home. There were about sixty of us in all—the women had been taken somewhere else; no one knew where. Everyone seemed mildly amazed to be there, but no one was too freaked out, or at least they weren't saying so. I tucked my shoes under my head for a pillow, did my best to sleep, and failed.

My mind was running fast: things I needed to tweet or blog, messages to get to the media. The oddest, most disconcerting thing about jail was being cut off from the flow of information, silenced. But it was also liberating—I'd spent the past two months in overdrive, endless conference calls, obsessing over messaging and framing, and . . . now I couldn't do anything. For the moment my only use was as one of a few dozen symbols, sitting behind bars. I couldn't help with what were clearly going to be crucial decisions: most important, if another wave of people showed up for the evening training, would we send them out to the White House knowing now that it might mean jail? The rest of the crew would have to figure it all out. I could relax, in a way I hadn't for weeks, and wouldn't for months to come. My body was uncomfortable—there really is no way to curl up on a steel slab that doesn't leave you bruised—but my mind was oddly at ease.

No one had a watch, and when we asked the guards what

time it was they enjoyed messing with us, giving answers hours apart. As far as we could tell, mealtimes were about three a.m. and three p.m., when someone shuffled down the hall handing out a bologna sandwich and filling your cup with a few inches of water if you held it out between the bars. ("Feeding time at the zoo," the guard would shout.) Eventually Saturday's adrenaline fully drained away, and I was tired, though falling asleep on steel was tough. After a while I woke up, sure that it was Sunday morning. Since I was feeling at least a little guilty (this was, after all, my idea), I was doing my best to keep people's spirits up. I knew there weren't that many churchgoers on hand, but people seemed to enjoy singing the old civil rights call-and-response hymn "Certainly Lord." ("Have you been to the jailhouse? Certainly, certainly, certainly Lord.") I asked Gus Speth to shout out a half-hour synopsis of his next book to the whole cell block, and when that was done—well, not much.

The guards had laughed when we asked them if we could see our lawyer. "Not on the weekend," one explained, which isn't exactly what the Constitution specifies, I think. So we had no idea of what they were charging us with, nor any way of knowing what was going on outside. Or, we wouldn't have, except for our secret weapon. And here I need to say a bit more about my next-door cell mate.

Gus Speth was a veteran of the first Earth Day in 1970. He'd helped found the Natural Resources Defense Council, then had gone on to head President Carter's Council on Environmental Quality, before running the United Nations Development Program and serving as a dean at Yale. At seventy he'd "retired" to Vermont but was now a faculty member at the state's law school. Oh, and his son was a high-powered D.C.

corporate attorney, who, it turned out, had been papering the jailhouse with one writ after another to try to find out what was going on with his dad. So it didn't completely surprise me when word filtered down the cell block midafternoon that Gus would indeed be meeting with his lawyer, weekend be damned. It was our sole connection to the outside world—what message should he send out to the media? This was my only chance to write so I set my mind to making suggestions, not that Gus really needed the help. Still, we were pretty pleased with what we eventually came up with. His message to the press read in its entirety: "I've held a lot of important positions in this town, but none seem as important as this one."

I held my breath while Gus was off in the interview cell with his son, waiting for the news—which, as it turned out, couldn't have been better. Day two of the protest had been, he reported with a smile, even bigger than day one, with nearly a hundred people getting arrested. The logic of the police, which probably works with criminals, seemed to have had the reverse effect on idealists: faced with more serious consequences, more people had come to the fore.

And with that I knew that the weekend really was a turning point: the moment when establishment, insider environmentalism found itself a little overtaken by grassroots power. We'd gone beyond education to resistance; the movement we'd long needed, and that had been glimpsed in environmental justice efforts around the country, was starting to emerge on a national scale. The spirit of that long-ago first Earth Day, when I was nine and Gus was a young man, seemed renewed—it was, I hoped, a portent of larger hope for the movement, a sign that people were finally willing to step up.

Since no one new was appearing in our cell block, we

guessed (correctly as it turned out) that we'd filled the jailhouse and the police had given up on their deterrence campaign and started treating the next waves of protesters in the usual way, with a trip to the station house and a hundred-dollar fine. With that thought I relaxed, and I think that second night I got a few hours sleep before the three a.m. bologna call. And since it was now Monday morning, the legal mill could resume its grinding—at five a.m. we were cuffed again and back in the wagon, bound just a few hundred yards for the holding cells at the city's central courthouse. The new wrinkle was that they shackled us to one another at the ankle—a chain gang, again just like the movies. Put your hand on the shoulder of the guy in front and shuffle along. Also, we got to meet many of the weekend's other criminals, some of whom seemed to find it worth a wry chuckle that we were there on the charge of "failure to yield."

"Sheeit," explained one veteran of the system. "That's not even a misdemeanor. That's a traffic."

We knew we were going to get out eventually. But it was sobering to see the justice system from the inside—to watch, for instance, as court-appointed lawyers appeared at the side of the cage, clutching a folder and bellowing the last name of their next assignment: "McClendon." Someone would shuffle sleepily toward the bars. "Attempted murder," the lawyer would say. A minute or two of conversation would ensue, and then the lawyer would wander off.

We watched this drama all day long, about twelve hours of lying on the cement floor or standing and holding the bars. No food. Finally, late in the afternoon, we saw the women from our team emerge from behind a door, waving at us as they were led to the exits. Half an hour after that we were let out of the cell, one by one, uncuffed, and then sent down the hall, through one more set of barred doors and out into the world, where there was fresh air, boxes of bagels on the courthouse steps, and lots of friends. And of course cell phones, with reporters on the other end. Back to work. On message. We won't talk much about the jail, we'd decided. Keep the focus on the Keystone pipeline. By the second or third interview I had my talking points: "People are willing to get arrested because this is a big deal. The second biggest pool of carbon on earth. Our numbers are growing."

And by nightfall, after a shower, a good meal, and—best of all—after I'd brushed my teeth, I could feel the jailhouse falling away behind me. We'd gotten though okay. It was a marker, a way to rally people—not the end of the world. The end of the world is the end of the world, which is why we'd sat down in the first place.

One other thing made me feel like I was back in the real world, of course: plugging into the Web. I read lots of congratulatory and sympathetic e-mails, and the requisite list of angry and abusive ones, but I also took a moment to glance at the news sites—which were starting to carry ominous stories about a tropical storm forming in the Atlantic. On the morning we were arrested the National Hurricane Center had reported that a cyclone was forming in the Lesser Antilles; by that night, as we were settling down in jail, its winds were strong enough that they had bestowed upon it a name: Irene. By late Monday afternoon, as we got out of jail, it had strengthened to near hurricane force and was clobbering Puerto Rico.

At that point, with forecast maps showing the storm

heading up the East Coast, it was time for cable news to go to DEFCON 1, with all the usual array of weathercasters standing on beaches pointing at surfers on waves and describing the carnage that might come in ninety-six hours: the usual run on Lowe's and Home Depot for plywood to cover windows, the usual stories about people who were going "to ride the storm out," the usual stories about mayors mad at people who were going to "ride the storm out," the usual over-the-top graphics. We didn't pay much attention to the forecast because we were too busy getting folks arrested each morning, but as the week wore on it became clear that the coming Sunday—halfway through our planned two-week siege of civil disobedience—would be the critical moment. The U.S. Park Police had already been asking if we'd suspend things that day because they needed all their officers to patrol the opening of the new Martin Luther King Jr. Memorial. Suspending civil disobedience to honor the man who brought it to these shores seemed odd, but the impending storm gave us a reasonable excuse to be nice to the cops, who were, after all, kind of our partners in this arrest business. So we agreed.

Eventually organizers called off the King ceremonies, too—the coming storm seemed too scary. As it turned out, though, Irene mostly skipped D.C. We had a windy party at St. Stephen's Church, with my friend DJ Spooky spinning records and telling stories about his trip to the Antarctic; outside the wind didn't exactly howl, but some tree limbs did fall, and the lights flickered once or twice. The same less-than-apocalyptic conditions prevailed northward to New York, where commentators had been awaiting a particularly juicy landfall only to be disappointed by a windy rainstorm. The media critic Howard Kurtz, for instance, offered a rant about how "the relentless tsunami of hype on this story" had been a sick joke "now that the apocalypse that cable television had been trumpeting had failed to materialize." Google, if you want, the words "Irene" and "fizzle"; it goes on for pages, largely climate-change deniers filling blog posts with sermons along the lines of "if they can't even forecast a hurricane, how can they tell us what the weather's going to be like in fifty years?"

I called home that afternoon to Vermont, and the neighbors said, "It's raining hard." I called back again that night, and they said, "It's still raining hard." Long after everyone at the cable news stations had stopped paying attention, Irene became a threat to my valley, maybe the greatest in its history. When Irene crossed open ocean off New York and New Jersey, the seawater was warmer than had ever been measured, and that let the storm clouds soak up great quantities of moisture; its winds hadn't done that much damage, but when it dropped that rain on the steep, narrow valleys of Vermont, all hell broke loose. Our house was high enough on the slope that it didn't suffer too badly—the power was out for days, but that's not entirely unusual. (We'd actually been hit far harder by record thunderstorms a few years earlier, whose devastation I described in my previous book *Eaarth*.) The worst rain was coming down just a few miles to the east, in the valleys of the Mad and White Rivers, which are isolated to begin with. There were no TV crews, no manic weathermen; while the media critics along the coast were debating whether the storm had been overplayed, the water was cascading down small streams and turning them into rivers; the rivers along the valley floors were raging as no one had ever seen them rage before. No TV crews arrived to document the destruction, but there were a few folks with cell phones—if you go to YouTube, you can watch 150-year-old covered bridges washing away in a matter of seconds. Those bridges had stood there since Abraham Lincoln's time, patiently taking everything nature could throw at them. But this was not the old nature—this was the new one we'd unleashed, that hybrid of natural and unnatural that is the distinctive mark of our time. More than eleven inches of rain fell in Mendon, Vermont, the greatest one-day rainfall in the state's long history. In an average year, Mendon gets thirty-seven inches of rain, which, in turn, is exactly the average for the whole United States. But that day it got a third of it all at once. That's what can happen now—it's what *does* happen, almost every day someplace around the world. Warm air holds more water vapor than cold—the atmosphere is 5 percent moister than it was when I was born. We've left the Holocene, and we've loaded the dice for both drought and flood, and on this particular day it happened to be Vermont that crapped out.

I could glean pretty quickly from the Web what was going on—I knew every stream now out of its banks, every bridge now washed away. Not going straight home was almost the hardest thing I'd ever done, far harder than going to jail, because I knew Vermont well enough to know exactly what would happen after the storm.

David Goodman described the events in the foreword to his small book *When the River Rose*, about Irene in the town of Waterbury, thirty miles north of my home. There Thatcher Brook and the Winooski River overtopped their banks, inundating dozens of homes, not to mention the state's psychiatric hospital. But never fear: "Vermonters were soon streaming into our community to help—the streets were so clogged with traffic that town officials went on the radio to ask the army of volunteers to park outside of the downtown. I ran into a dozen members of the Green Mountain Club Trail Patrol who came down from the mountains to help dig out homes on Main Street." The CEO of Ben and Jerry's was mucking out the basement next door. At the elementary school, Marni Martens had volunteered for the task of greeting other volunteers as they arrived. Partway through the afternoon the second day, she reported, there was a momentary lull in the flow of new shovelers. "I was left with one homeowner standing alone in tears because there were no volunteers left," she recalled. "Here was Sonja—a mom, a parent of kids I teach—crying in my arms because she just couldn't do it herself. I told her we would get volunteers. So I did what anyone in Waterbury would do: I ran down the street to WDEV and asked them to announce that we still needed volunteers at the school." Right away two large groups phoned to say they were on their way.

WDEV by the way, is one of Vermont's greatest resources. I wrote an essay about it once for *Harper's Magazine*—it's a true community radio station that carries the Red Sox and the stock car races, and has its own in-house bluegrass band (the Radio Rangers). And in times of crisis—well, everyone at the station, from the ad salesmen to the owner Ken Squier, heads out to find out what's happening so they can tell their neighbors. The Internet was no use those first days, because the power was gone; but if you had a battery in your transistor radio, you were in touch with WDEV, except for the few moments when the station was shut down in order to pour more gas into the emergency generator. Announcer Eric Michaels was up all night as the water was rising, taking cell phone calls from people in every kind of emergency. "I read stories of sheds passing by in the floodwaters," he said. "I read texts that cows were floating past. And cars. And homes. And caskets ripped from flooded graveyards."

In other towns, newspapers became the organizing principle—Vermont still has some fine local papers—relics, like WDEV, of another age (or maybe precursors of the next one). In Randolph, at the northern end of the next valley over, it was the *Herald* and its editor, Dickey Drysdale, who told the stories of men on ATVs catching bags of prescriptions thrown across the canyon at the foot of Camp Brook Road and somehow making it over the Bethel Mountain Road to the cutoff village of Rochester. Or of the Mighty Mud Brigade, thirty strong, going from house to house with shovels and buckets till everyone was mucked out. The *Herald* had great pictures, too, including one of a woman standing next to her wrecked farmhouse with a huge sign that just said, "Irene You Bitch."

It's not like this huge effort lasted a few days—it went on for weeks and months, as volunteers and government officials did all they could to put the state back together. It wasn't always pretty: the governor waived various environmental laws, allowing bulldozers into rivers, and in the process doing damage that will probably make the aftermath of the next storm worse. But his highway department worked at a feverish pace—and though it took till January 2012, every mile of state road eventually reopened, and the crucial bridges were replaced. (Replaced with longer spans, in the expensive hope they'll deal with larger floods.) The state legislature spent the next session figuring out how to account for it all, with fingers crossed that the next disaster would wait till the bonds were paid off. Plenty of people will never completely recover—as usual, the poorest were the hardest hit, because there were a lot of trailer parks on flood plains. But I think David Goodman put it best at the end of his slim book: "The raging rivers of Irene revealed something more beautiful and durable than the wood and steel it tore away: the incredibly generous spirit of Vermonters, and the ties that bind our communities."

The ultimate triumph of community over crisis: it is as hopeful a note as I know how to strike. When people ask me where they should move to be safe from climate change, I always tell them anyplace with a strong community. Neighbors were optional the past fifty years, but they'll be essential in the decades to come.

Still, there comes a point past which neighbors are no longer sufficient, a point we are fast approaching. What if it had rained fifteen inches on Mendon instead of eleven? Across the United States, in the five decades I've been alive, the number of extreme downpours a year has increased 30 percent. Across New England, it's gone up 85 percent. Across Vermont, it's literally doubled. The intensity of the largest rainstorm each year has grown by more than a fifth.

And all of this is with one degree of temperature increase.

The same climatologists who told us that this would happen now tell us we're likely to see temperatures rise *five degrees* this century unless we get off coal and gas and oil far faster than any government now plans. There's no adapting to that, not even in relatively rich places such as America. And in poor places? In Pakistan, a year before Irene, flooding forced twenty million people from their homes. Twenty million people is thirty-five times the population of Vermont.

So that's why I stayed in Washington. I was constantly on the phone with family and friends. But I wasn't shoveling out basements, I was shoveling folks off to jail. And for me the most moving moment of the whole long siege came the following Tuesday morning, when I got to Lafayette Square and found a busload of Vermonters lined up waiting to get arrested. They told me the story of their trip—an odyssey of driving endlessly to find intact bridges and roads so they could get out of Vermont and down to D.C. "It's too late to stop Irene," one woman said. "Maybe it's not too late to stop the next one."

It didn't move me just because these were my neighbors. It was more because they helped me bridge, with their trek, the gulf in my mind between home and away—that sense that the halves of my life didn't connect. I described earlier how much I wanted to be home, free from the ceaseless wandering, building the kind of local economies the planet badly needed. But in the rains of Irene, the contradictions seemed to dissolve and melt. The global implied the local and vice versa, and it was my particular fate to straddle the two.

Though I trust Robert Frost on most things, these two paths were converging. If we couldn't win this fight at the global level—if we couldn't bring climate change under some kind of control—there was no use even trying to make local economies work. So far we've only raised the temperature a single degree, and yet that's been enough to melt the Arctic. And the same scientists who told us that this would happen are firm in their consensus that unless we get off coal and oil and gas with great haste, that one degree will be four or five: enough to turn farm belts into deserts, make humid New England a swamp, and reduce Irene to just another link in an endless chain of disasters that will turn civilization into a never-ending emergency response drill. They don't call it *global* warming for nothing. We *have* to work at the scale of the planet.

At the same time, we've already changed that planet. And no matter how well campaigners like me do our job, we are going to change it some more. Climatologists insist that even if we stopped burning fossil fuels tomorrow, the temperature and the damage would increase for decades to come. So we need new local economies if for no other reason than that they'll weather the coming storms a little better. In place of our too-big-to-fail systems of banking and energy and agriculture, we need squat, hardy, scaled-down versions. Small enough to succeed.

In that second and final week, the protests just kept growing. Every morning, in the strangest commute of my life, I'd walk down Connecticut Avenue to Lafayette Square and give much the same short talk: "You're very brave. It's hard for normal people not to move when a policeman says 'Move.' We're not built that way—it feels uncomfortable, psychologically uncomfortable. You're brave, really brave." And they were, day after day. Some days a busload would arrive from one of the states along the pipeline route: Nebraskans, Texans. Some days our friends at the Indigenous Environmental Network were in charge—sage burning, drums pounding. A few movie stars took part: Margot Kidder, who'd played Lois Lane, and was from northern Canada. Tantoo Cardinal, a Cree actress born in Fort McMurray in the heart of the tar sands—you saw her in *Dances with Wolves*. Darryl Hannah.

But participants were mostly just ordinary people who had decided to do the extraordinary, arriving from every state in the union.

The last day, while the arrests were under way, we held a rally in Lafayette Square. A band played, and a hundred people carrying a four-hundred-foot-long inflatable black pipeline snake-danced through the area—it was a three-ring circus that went on for hours because more than 200 people insisted on getting arrested. They brought the two-week total to 1,253, the biggest civil disobedience action since the protests against nuclear testing in the 1980s—even much bigger events, including the protests in Seattle against the World Trade Organization, hadn't yielded as many arrestees. And not perhaps since the civil rights movement had there been a crowd of protesters quite as diverse.

We'd asked, you'll recall, that college students not have to bear the burden of this particular fight. And people had taken the message to heart. We didn't actually ask how old people were, since that would have been rude, but we did ask who was president when they were born, and the largest cohorts had been babies in the Truman and FDR administrations. Not many things get easier as you age, but getting arrested is one—an arrest record for a twenty-one-year-old might cost you a job, but past a certain age, what are they going to do to you? On the last day, eighty-six-year-old Roy Ingham was loaded into the paddy wagon with a sign around his neck, "World War II Vet, handle with care." We'd also asked people to show up wearing their Sunday best, and that's what happened—lots of sundresses, neckties, sports jackets. It wasn't because we were a formal bunch, but because, as I'd said in that initial letter, we wanted to send a visual signal to everyone looking on: *There's nothing radical about what we're doing here*. We're just Americans, interested in preserving a country and a planet that looks and feels something like the ones we were born on.

Radicals? They work at oil companies and coal companies and gas companies. They're willing to alter the chemical composition of the atmosphere to make money. No one has ever done anything more radical than that.

As the sit-in swelled, and as the logistics became a little more routine (after a while, organizing crime is just another job), we started to think about what would happen next. We hadn't given it much thought before—after all, if the civil disobedience had fizzled, there wouldn't have been any real campaign.

But now—well, now 1,253 people had anted up with their bodies to get us in the game. And it had worked. We hadn't gotten much press coverage, but enough—for those paying attention to environmental issues or presidential politics, Keystone was now on the mental map. I confess we were a little stunned at how well it had gone—and, having bought our chips, it was time to play our hand.

Official Washington was wired to approve the pipeline. The State Department was conducting the official review, and Hillary Clinton had said even before it began that she was "inclined" to approve the permit. There was a good reason for that (the Canadians wanted it badly, and they're our closest ally), but there were also several bad reasons—as Freedom of Information Act requests were beginning to show, lobbyists from the pipeline's builders, TransCanada, had successfully infiltrated the department, paying heavily for the lobbying services of many of Clinton's biggest campaign donation "bundlers." Everyone who followed this kind of thing said our case was hopeless—indeed, when the Capitol Hill magazine *National Journal* polled three hundred energy insiders that fall, 91 percent of them said the permit would be granted. Bad odds.

But we had one card to play. President Obama would make the final decision; before Keystone could be built, as I've said, he had to declare that the pipeline was "in the national interest." So reaching him would be the goal, and we figured the next few months were our only opening. His reelection campaign was gearing up; he was heading out for more speeches on campuses, trying to rekindle some of the enthusiasm of 2008. Weighed down by the economy, he was trailing in the polls and his advisers feared that young people and liberal donors were tired of his compromising ways. In fact, two days before our sit-in ended, the president unexpectedly turned down a proposed EPA smog rule, a regulation to curb dirty air that even George W. Bush had supported. I cursed when I saw that news—but I confess it also occurred to me that it might help our effort. He'd given environmentalists precious little, and now he might feel the need to give us something.

So before we left Washington, I thought we should talk to the White House. The administration hadn't reacted to our sit-in—the president had been on vacation in Martha's Vineyard for the first half of it. Jay Carney, the president's spokesman, told reporters that he "hadn't told" Obama what we were up to. But we figured someone must have noticed, and indeed when my colleague Jamie Henn called the administration's Office of Public Engagement to arrange a meeting, we were quickly offered a time the next morning, the second to last day of our protest. It was to be private and unofficial—we suggested the bar at the venerable Hay-Adams Hotel, because it was right next to Lafayette Square. And because, I confess, it appealed to my sense of the absurd.

The Hay-Adams is named for the two men who lived on the site before the hotel was founded: William McKinley's secretary of state John Hay, and Henry Adams, great-grandson of John and grandson of John Quincy and our finest nineteenth-century writer about American politics. It's a power place—Obama and his family stayed there the two weeks before his inauguration in 2009. And it shares the block with the U.S. Chamber of Commerce, the biggest donor to American political campaigns, outspending the Democratic and Republican National Committees combined. (And, not coincidentally, the most important opponent of action on climate change, going so far as to file a brief with the Environmental Protection Agency explaining that global warming was not worrisome because humans would "adapt their physiology" to deal with a warming world.) In fact, Jamie went off to the bar to fetch a round of Diet Cokes, and he came back with news that Tom Donohue, the head of the chamber and arguably the second most powerful man in D.C., was two tables away having a tête-à-tête of his own.

Jon Carson was our guest. In his midthirties, he'd ended up in the White House after a career as a political operative and a stint in the Peace Corps. In 2008 he'd served as field director of the Obama effort, perhaps the greatest and most interesting presidential campaign ever. Now he ran the Office of Public Engagement, in charge of—well, public engagement. "We create and coordinate opportunities for direct dialogue between the Obama Administration and the American public, while bringing new voices to the table and ensuring that everyone can participate and inform the work of the President." That's how the White House Web site explains it. But politics surely was part of Carson's brief, and my guess from his biography was that he'd respect the scale of what we'd pulled off the last two weeks.

Not that I gave him a chance to tell us, one way or the other. I began by explaining that good Gandhian practice demanded that we tell our adversary our plans in advance, but truth be told I wasn't feeling Gandhian at all. I was doing my best instead to channel Leo McGarry, the chief of staff from the television series *The West Wing*. Powerful people intimidate me; my instinct is to try and make friends. So McGarry in steely mode is my model. Don't shout, but speak firmly and as if you have all the power you need. I didn't expect Carson to give in to us, and indeed I barely talked about the pipeline at all; but I did need him to pay attention—to understand that we weren't normal Washington lobbyists who would be appeased by half measures. I needed, I thought, to make an impression.

So for twenty minutes of unbroken eye contact I just talked. Here's what we were going to do: Follow the president wherever he went, making sure that the pipeline message became ubiquitous. And we were going to round up every big donor we could think of, using them to get the message back to the White House that this pipeline was now the measure of his environmental commitment—that it was the first environmental cause in thirty years that had filled the jails.

More to the point, though, was what we *weren't* going to do. "We won't do you guys the favor of attacking the president," I explained. "If we do, we're angry extremists, easy for you to marginalize. Instead we're going to pay you the much more dangerous compliment of taking your words seriously." I showed Carson some of the president's quotes from the campaign he'd run. "It's time to end the tyranny of oil," for instance. Or, most powerfully, the close of the speech Obama had given the night he clinched the nomination.

> I am absolutely certain that generations from now, we will be able to look back and tell our children that this was the moment when we began to provide care for the sick and good jobs to the jobless; this was the moment when the rise of the oceans began to slow and our planet began to heal; this was the moment when we ended a war and secured our nation and restored our image as the last, best hope on earth.

"He shouldn't say stuff like that if he doesn't mean it," I said. "If this is the moment when the planet is going to start to heal, you don't get to tap the tar sands. We will say a thousand times in the next few months that we're confident the president will do the right thing, because he said he would. We're going to refuse to be cynical. We're going to be deliberately naive. And if, in the face of that, you approve the pipeline, you're going to take a hit. You'll have to decide if the hit is worth the grief you'll get from the oil industry. You're the political professionals—you'll be able to make the calculation. But you can count on us to do our part."

We shook hands and left, heading back to Lafayette Square where the police were busy hauling away that day's protesters. I had no idea if we could keep our promise.

• • •

Perhaps I've given the impression that I'm a courageous fellow, ready to trot off to jail at the drop of a hat, able to stare down presidential aides. That would be not quite true.

In fact, on the long list of things that scare me, bee stings rank fairly high. And with some reason. Once, wandering the woods a mile or so behind the house where I lived for many years in the wild Adirondacks, I stepped on a yellow jacket nest. As usual I was lost in some reverie, when all of a sudden a drench of incandescent pain splashed up my stomach toward my head. It came so fast—a wash of pure feeling, as if someone had tossed a pot of boiling water at me—and it hurt so much, a distillation of pain I've never experienced before or since.

It took several seconds to figure out what on earth was happening. I'd been climbing a steep slope, pulling myself up hand over hand one sapling at a time, which is why the yellow jackets were at waist level when they boiled out of the ground. I couldn't see them—I just turned and ran, pure instinct demonstrating its limits because running on a forty-degree slope doesn't really work. I cracked my head on a tree branch, cutting my forehead and closing my right eye. But the pain was still there, and I kept running—after a few hundred yards of nearly blind flight I was able to flick the last insects off my neck and pull off my shirt.

The shriek of pain subsided to a dull roar, enough that I could collect my wits (though not my eyeglasses, which were somewhere Back There where I was definitely not going). I knew the woods so well that I was able to make it back home on autopilot, slightly panicky because big red hives were swelling across my torso. I thought of all the stories I'd read about shark attacks or lightning strikes that ended with the suddenly less comforting statistic that "more people die each year from bee stings."

I didn't perish; I made it home, and my scared wife drove me at top speed the forty miles to the hospital, where they hooked me up to six kinds of IV and counted my stings (seventy-six). And that was that. Not entirely—I've written elsewhere that the moment was a kind of epiphany, when the scrim between me and the rest of the natural world thinned, when, for a few hours that stretched into a few weeks and has never entirely gone away, it felt like I was part of the natural world around me, not just an astronaut moving through it in my bubble of thoughts and plans. So, that was compensation.

But the experience also left me really scared of getting stung again. The doctors had warned that I was now likely far more sensitive to yellow jacket venom, and that if it happened again my throat might very well swell up and I'd get to contemplate my intimate connection to the natural world in an entirely different way. My GP gave me a prescription for an EpiPen, so I could jab it into my thigh if lightning struck twice. In fact, given my predilection for wandering far and wide, he made me carry two of them, so I'd have some chance of getting back to the road. I've been stung since, but only by one insect at a time. I've swelled profoundly each time but never so badly that I had to use my spike. Still, I can perhaps be forgiven for being just the littlest bit wary of Kirk's bee-yards with their buzzing clouds. Even though he assured me over and over that honeybees were different from yellow jackets, that their venom would probably be harmless, and that some people got stung on purpose for their arthritis,

Kirk finally figured out I was a baby and gave me a bee suit to wear.

He had one, too, but he often worked without the veil, or didn't pull on the gloves. My routine never varied—a safe fifty yards from the hive I'd pull the white coveralls over my clothes, put on the special white gaiters that kept bees from flying up my pant cuffs, placed the helmet on my head and carefully tucked the veil into my collar, and finally pulled on the long white gauntleted gloves. And then I felt more or less safe, though the first few times I worked with Kirk I know I was still tensing my body and hunching my shoulders.

Happily, working with him (which mostly involved carrying crates of bees to and from the truck, and fairly often involved staying out of the way) was fascinating—it was like getting to be the clueless studio assistant to Caravaggio, with an up-close glimpse of a master at work: a master who, happily, was willing to explain what he was doing as we carried crates. I slowly pieced together an understanding of just how innovatively this solitary and shy man had reinvented his profession. I'd known he was "chemical-free" and that this was "good," but I hadn't understood that it was only the most recent of the changes he'd pioneered.

By chance, the day Kirk first gave me the bee suit we visited one of his biggest beeyards, just down the road from the house he'd rented for many years before we found the new farm. "This is really where it began," he said. "This is really the place where I figured out that I could make my apiary work."

His revelation grew out of an accident in the winter of 1988. Long tradition held that you needed a big colony to survive a northern winter—at least two big wooden boxes of bees. But "one year I had some smaller colonies left over, two in each box," he recalled. "I didn't remember to combine them, I just left them on top of some other boxes. When I eventually noticed, I said to myself, 'They're done for,' but when I came back in the spring they were all alive. The next year I did eight in that way, and they all survived. That spring, when I unpacked those colonies and found them living, that's when the whole thing fell into place in my mind. I could see it all: raising new queens in midsummer, keeping them over the winter in nucleus colonies, and having this great new product to sell." He'd changed the math in a decisive way—one that presaged real profits.

From the early twentieth century till that day, northern apiarists had mostly purchased new queens each spring from breeders in the south. "I'd always wanted to raise queens from the start, but I figured there must be some reason that no one did it," said Kirk. "You'd get them by mail, that was how it was. The man who was the dean of northern apiculturalists then, a fellow at Cornell named Roger Moore, just flatly said, 'Nucleus colonies don't survive winter in this latitude. Take them to Florida.' He wouldn't even come to the beeyards to look at what I was doing. I remember once going to give a talk at the New York State beekeepers meeting and when I got up to speak he got up to go judge the honey entries."

Overwintering small colonies meant you could multiply hives fast enough to make some real money—suddenly a beeyard that could yield $3,000 worth of honey could also produce enough queens to be worth five times that much. And it meant Kirk could start working on the genetics of his now largely self-contained apiary; instead of importing queens from someone else, he could pick them from the very best of his colonies and propagate them rapidly—which allowed him, years later, to stay barely ahead of the mites and pests that wrecked so many other beekeepers. "I was the first one to make it a practical system since the early 1900s," he said. "I'm proud of that."

But we'll get to mites and chemicals and so on a little later. What I liked from the very beginning about working with Kirk is that alongside his pragmatism, he thought philosophically about bees as well. He waxed romantic, almost. Because this beast he was busy domesticating was also, simultaneously, very wild, going forth every hour to explore the wider world. "It's kind of like fishing for me," he said. "The resource out there, the fields, it's just a kind of ocean. It's like being there putting your nets out. And at its best it's just remarkable. You know how the Bible talks about a land of milk and honey? Well, they're very closely related. The same leguminous plants that are great for cows—the clovers—are great for bees, too. But the clover doesn't come up strong every year. That depends on the weather. The honey either materializes in the hives or it doesn't, and you often don't have that much control."

3

. . . . .

HONEYBEES AND CONGRESSMEN

Civil disobedience is hard work—training a hundred new people every night, getting them safely arrested in the morning, writing press releases, talking to reporters. In the downtime of the afternoon hours we were doing our best to rally the rest of the environmental movement to this upstart effort. Which was key, because the "environmental movement" had largely become a collection of environmental groups, each doing impressive work but often without enough connection to the grass roots or to one another. (The grassroots groups, by contrast, were vital—but their voices didn't break through to the national level often enough.) Big Green's supporters had largely aged, and many were best at writing checks. We wanted to help these groups see this new surge not as a threat but as an infusion. We also needed their help to make Washington, a city accustomed to its elites, pay attention.

Fresh out of jail, I'd written a letter for the CEOs of the

Washington green groups to sign, and all of them did, even though the language was stronger than some were used to using: "Mr. President, there's not an inch of daylight between our position and those of the people getting arrested in front of your house." About ten days into the siege Al Gore released a two-sentence blog post pointing out that the tar sands were the "dirtiest source of liquid fuel on the planet," and saluting the people who had "bravely participated" in the arrests. On the one hand, this was expected—Al Gore was the greatest climate advocate on the planet. On the other hand, he's a loyal Democrat, who'd once hoped to sit in the White House we were now besieging. It felt like we were starting to break through just a little. It also felt like we were taking the movement new places: one night Benjamin Jealous, the young and dynamic head of the NAACP, showed up to address our training session. The crowd on hand, used to being told that environmentalism was something rich white people did, were thrilled by his tales of civil rights organizing; it was just as powerful when the Indigenous Environmental Network showed up in force and the even younger but no less dynamic Indian leader Gitz Crazyboy told tales of life in the tar sands desert.

And we would wake up each morning to news of what our allies from 350.org were doing overseas. In Cairo, delegations arrived to see the Canadian and American ambassadors. Our friends at 350 New Zealand shut down the Canadian embassy in Wellington for the afternoon, waving an oil-soaked maple leaf flag. Judging by the number of CBC journalists suddenly on the scene, the protests were coming as a bit of a shock to the Canadians, who had mostly either celebrated or overlooked the fact that their nation was trying to become the next Saudi Arabia. One morning the news was even less expected: ten of the most recent Nobel Peace laureates, led by Desmond Tutu and the Dalai Lama, had written a letter to Obama urging him to block the pipeline. Nonviolent protest was working in textbook fashion—moved by the courage on display, people were responding. Sometimes I'd see the members of our crew just looking at one another and grinning.

We knew we couldn't keep the arrests going indefinitely, though, and we knew that the Obama administration would shrug off pressure from the usual suspects. So we had to figure out what to do next—how to take the movement off Pennsylvania Avenue and into places like Pennsylvania.

Once the final arrestees on the final day had been hauled away in handcuffs, our small circle of core organizers sat down at a Thai restaurant to eat and plan. Somewhere in the middle of the pad thai, Matt got a text that the police had released the last of our band, and I felt a huge weight lift. Improbably, for two weeks and through 1,253 arrests, nothing had gone wrong. But now what?

We knew, in the broadest terms, that we had to carry through on that promise I'd made to that White House aide—that we'd need to fan out across the country for the next few months. The president, after all, had promised he'd make a decision on the pipeline by year's end. But the country is . . . large. Deprived of our focus on the White House, how would we keep momentum? So we knew we'd have to come back to D.C., too. Some pushed for a new round of arrests—even more civil disobedience. But my sense was that we were walking a fine line between pushing the president and pushing him against a wall—too much pressure and he wouldn't be able to give in, lest he look weak. Still, just coming back to Washington for a rally was likely to be anticlimatic—it had been decades since a "march on Washington" had really accomplished much. "What if we surrounded the White House with people?" I asked, mouth full of tom yum soup. It wasn't a great idea, perhaps, but it was an *idea*: something fresh enough that it might interest both supporters and reporters. Everyone was intrigued, but we made no decisions, except to head across the street to a club where hundreds were gathering for an after-party. Beer was the order of the evening, as were many toasts to our general fineness.

It turned out that we needn't have worried too much about surrendering our White House rallying spot. Not for the last time our luck held, and the president chose that very week to set out across the country in what amounted to the start of his reelection campaign. He headed to college campuses in swing states, confident of the same happy reception that had greeted him in 2008. And for the most part he got it—big crowds of kids eager to see the president. But everywhere he went he also found growing numbers with a new chant. "Yes We Can . . . Stop the Pipeline." It began at North Carolina State University in Raleigh—a blog post from a student captured the scene: "Never in all my years at NCSU has such a rally occurred—a goosebumpy moment." The same thing happened the next day in Columbus, Ohio, when six locals who'd been arrested in Washington gathered a large crowd to line the motorcade route. The president didn't go to Harvard that first week, but his campaign manager Jim Messina did—and he was greeted by twenty-five students who had volunteered for his campaign but now were chanting, "Obama Can Stop the Tar Sands." The blogger John Chandley was at the scene, and he watched Messina duck through a side door to avoid the peaceful demonstration. "That's your base, Jim," he wrote. "You're losing them. Time to listen up and pay attention." As the *Nation* put it in an editorial: "Candidates for president routinely make promises they don't keep. But voters aren't stupid. What matters is why a candidate breaks a promise: is it because he won't deliver, or he can't? Obama has 13 months to persuade voters that they should blame not him but the GOP for his presidency's shortcomings. He has much less time to convince the thousands of activists nationwide—who do the grunt work of getting out the vote—that he's worth their sweat and sacrifice one more time."

The *Nation* and Harvard are one thing. But the mood was spreading. Nebraska had long been a focus of opposition, because plans called for the pipeline to cross the state's cherished Sandhills and the Ogallala Aquifer. Jane Kleeb, the founder of Bold Nebraska, had been building a movement there for more than a year, strong enough that even the state's Republican governor and U.S. senator had come out in opposition to the Keystone project (a fact that would prove tactically crucial). Her efforts were reaching a crescendo in early September 2011. The state university's beloved Cornhuskers were playing Fresno State in Lincoln when a halftime video of highlights from the 1978 Big Eight football championship squad came up on the Jumbotron. At the end, the logo for the video's sponsor flashed on the giant screen—TransCanada, touting what it was calling the Husker Pipeline. As one reporter described the scene, "Tens of thousands of fans proceeded to swallow their beer, put down their food, and boo. It was actually more than booing. It was more like loudly seething. I don't think the Oklahoma Sooners ever produced a reaction like this." The next day Nebraska's athletic director (and former legendary coach and Republican congressman) Tom Osborne announced that the university was rescinding its $200,000 agreement with TransCanada: "We have certain principles regarding advertising in the stadium such as no alcohol, tobacco, or gambling advertisements. We also avoid ads of a political nature. Over the last two or three months, the pipeline issue has been increasingly politicized. Our athletic events are intended to entertain and unify our fan base by providing an experience that is not divisive." Nebraska is literally a red state—bright tomato Cornhusker red. I read the accounts in the Nebraska papers that Jane sent my way and I thought: "Huh. Maybe we've got a chance."

Not a great chance, of course. As the fall wore on, we learned an awful lot about how effectively TransCanada's lobbyists had done their job. The State Department was supposed to be conducting the official review, and documents obtained by Friends of the Earth showed not only that TransCanada had hired Hillary Clinton's former deputy campaign manager, Paul Elliott, as its chief lobbyist, but that his old colleagues were rooting him on. "Go Paul!" one high-level State Department staffer e-mailed her old friend whenever he sent out news that a senator had come out in support of the pipeline. As a *New York Times* account put it, "The exchanges provide a rare glimpse into how Washington works and the access familiarity can bring." One of the e-mails put it even more succinctly. Marja Verloop, an official at the U.S. embassy in Ottawa, consoled Elliott when he groused that environmentalists had been complaining about his efforts. "Sorry about the stomach pains," she wrote. "But at the end of the day it's precisely because you have connections that you're sought after and hired."

Indeed, TransCanada's connections were so good that the State Department allowed it to choose the company that would review the pipeline's environmental impact. TransCanada chose a company called Cardno ENTRIX, which boasted on the front page of its Web site that TransCanada was one of its "major clients." In effect, TransCanada was allowed to review itself—not surprisingly, it found that a project that NASA scientists said would speed us toward "game over for the climate" would have "minimal" environmental impact. This is how it works in Washington—we weren't naïfs, but I confess I was a little shocked by how blatant it all was. We'd picked this one project to focus on, and every closet we opened was a boneyard. And the more we protested, the more the people who knew how the game worked told us we would lose.

We didn't slacken the pace, though. Obama in Seattle? So were hundreds of protesters. Obama in San Francisco? Thousands, including some of the big Silicon Valley donors who'd backed his campaign in 2008. In Atlanta they picketed Obama campaign offices, and in Oregon they gathered outside fundraisers: "Hey Michelle—No XL," people chanted as the First Lady's motorcade arrived in Portland. And while all this was happening, the Occupy movement was growing fast across the nation. It had begun in New York exactly two weeks after we'd finished our Washington sit-in. I'd taken the train to Manhattan as soon as I could and spoke through the mighty human microphone, each ring of listeners repeating my words till they reached the edge of the throng. "Wall Street's been occupying the atmosphere for decades," I said. "It's about time we returned the favor." Before the fall was out I'd spoken at nine Occupy encampments across the country. (They were true to their roots. In San Luis Obispo, statistically America's happiest city, there was a man with a sign that read "Free Hugs." In Boulder the Occupiers were as handsome and fit as all their neighbors; it looked like they were shooting an Occupy catalog, and after my talk they gave me a gift bag of herbal teas.) Our messages synched easily, since the oil companies were the 1 percent of the 1 percent, and since the corruption around the approval process was a poster child for complaints about the unfairness of our political system. Journalists complained that Occupy didn't take positions, but in our case it simply wasn't true—the New York City General Assembly enthusiastically endorsed the fight against Keystone. Out west, meanwhile, ranchers and indigenous leaders were holding a session of their own on the Rosebud Sioux reservation, producing my favorite headline of the fall: "Cowboys, Indians Unite Against Pipeline." (Not politically correct, perhaps, but politically potent.) In Washington, D.C., "Tar Sands Students" formed—and a hundred of these high school leaders went to the State Department to meet with an assistant secretary of state—the picture of them standing sweetly in their good clothes on the Foggy Bottom sidewalk and clenching their fists for the camera made my day.

Every day there was something like that. Our crew was spread out across the country, but in constant digital touch. We'd open the e-mail and there'd be a picture from the Midwest—a young couple who'd spearheaded a drive to carve Obama jack-o'-lanterns during the 2008 campaign ("Yes We Carve") was now producing anti-Keystone pumpkins. Or there was a photo from St. Louis—a pair of young volunteers who'd gone to an Obama fund-raiser in their sharpest outfits, only to open their jacket and shawl when the president began to speak to reveal their anti-Keystone banners. The president didn't respond immediately, but during his remarks he said, "We've got a couple of people here who are concerned about the environment." We seized on remarks like that as proof that we were getting the president's attention, and we knew it for sure a couple of weeks later when the president gave a big talk in Denver. Speaking at the University of Colorado, he was greeted by the now-usual throng with an anti-pipeline banner. But he was also interrupted midspeech by Tom Poor Bear, vice president of the Oglala Lakota Nation. He told the president to stop Keystone, or at least he tried to—security guards hauled him away before he could get very far, so we did our best to spread his statement: "As our great leader Crazy Horse once said, 'You cannot sell the land your people are buried on.' I believe today he would say, 'You cannot desecrate the land your people are buried on.'"

Before the hour was out I got a call from a steamed White House aide explaining that it was not okay to interrupt the president. I was surprised, actually, at his real anger—at the idea that we'd committed some act of lèse-majesté. I could truthfully assure him that while we hadn't told Tom what to yell, we weren't going to tell him to shut up. (It seemed to me as if American history had earned the Sioux the right to say a word or two.) For the most part, though, we kept true to our plan of not attacking the president—a plan that reached its zenith on November 6, 2011.

That was the day—precisely one year before the election—that we returned to Washington, intent on encircling the White House. I'd convinced my colleagues that it was a decent plan and then they'd done all the work, figuring out the logistical challenge of surrounding the mile and a half of security perimeter around the executive mansion. (Depending on your militancy level, we said we were either putting Obama under house arrest, or giving him an O-shaped hug.) The day dawned perfectly blue, and with all kinds of good omens: the *Washington Post* had a piece demolishing the industry claim that the pipeline would create jobs, a Julia Louis-Dreyfus video recruiting protesters was going viral, and the three thousand bright orange safety vests we'd ordered had arrived just in time. Three thousand people was what we were hoping for—close scrutiny of Google Earth had convinced our crew that, hands outstretched, that many folks would make it all the way around. As it turned out, we needn't have worried—such a crush of people descended on Lafayette Square for the rally that we immediately knew our problem would be to deploy them without knots and jams. As speaker after speaker gave crisp calls to action, we divided the crowd into fourths and then sent them to each side of the perimeter fence, following giant colored flags. Within half an hour, the word from our observers on every edge was that the White House was fully ringed, and that most of the way people were standing shoulder to shoulder five deep.

Michael Brune, the director of the Sierra Club, grabbed my arm and said, "Let's do a lap," and it was the most fun half hour of the whole fall, high-fives nonstop till my hand was swollen. It had been, in fact, a nonstop autumn (someone on our team calculated at some point that I'd spent more nights in jail than I had in Vermont), but we'd done what we could. If there was ever a happy protest, this was it—every banner was a quotation from President Obama, every chant a hopeful call for him to act. We'd played our hand as best we could.

But how would we keep the pressure on? We sat up late that night, hammering out a tentative plan. If we didn't hear from the president we'd have to go a little harder, this time visiting campaign offices in all fifty states. But we knew it would be tough to keep the positive vibe, we knew that our tone would start to slip. We knew we risked pushing in a way that would make it impossible for Obama to give in.

So relief hardly begins to describe my feeling two days later when a call from the White House alerted me that "something's in the works—something you're going to like." The relief still came with some tension. We knew that the president wasn't going to veto the pipeline outright—that the most we could hope for was that he'd delay the decision for a year. And we knew he wouldn't cite our pressure. From the beginning we'd encouraged the White House to grab the lifeline that the Nebraska Republicans had given him, a scant cover to pretend this was a bipartisan action. But we also needed more than that. If the only issue he cited was possible spills as the pipe crossed the Sandhills, then TransCanada would be able to chart a new course and the pipeline would certainly be built. To have a chance in the long run, we needed the president to promise that when he made the final determination, which now seemed likely to come no earlier than 2013, that he'd take climate change into account.

That was the bargain I kept pushing as the week wore on and I wandered across New Mexico, in and out of cell range. I'd speak to students at United World College and then talk to our White House contacts; one crucial e-mail came while I was at Occupy Santa Fe—as stereotypically beautiful as you might expect, with a full moon rising over a campfire. On Wednesday night I spoke to a standing-room-only crowd at the gorgeous old Lensic Theater in downtown Santa Fe; the next morning I was at the Albuquerque airport, and I knew the news was starting to leak because my in-box was filling with calls from reporters. First the State Department and then the White House put out their releases: the review would last another year, and in the end a number of factors would be considered, including "climate." So.

As I told reporter after reporter, environmentalists never win permanent victories, and this was obviously no exception. It wasn't exactly what we wanted—but damn. I kept thinking of that *National Journal* poll with 91 percent of those in the know certain that the permit would go through. And I kept thinking of the huge piles of steel pipe TransCanada had already prepositioned in fields across the Midwest—hell, they were so confident they'd *mowed* the whole pipeline strip a few weeks before so they'd be ready to go the second the president gave them what they'd paid for.

But he didn't. The pipeline might get built someday, but not that day. We hadn't stopped global warming. But for once we'd shown that people could actually stand up to the richest industry the world has ever known. I answered as many phone calls as I could and then sunk into my seat on the airplane and slept. For the first time since those nights in jail I wasn't tense—wasn't trying to figure out the next move.

For about a week I managed to maintain the illusion that the worst of the fight was over—that we'd really have a year of quiet on the Keystone front.

I went home for the first time in weeks. People always say it's a pleasure to sleep in your own bed, which is true, but it's also a pleasure *not to* dispense your Raisin Bran from a plastic cylinder by turning a knob. I've learned to tolerate the

Courtyard by Marriott and the Homewood Suites and the Days Inn—if there's an Internet connection then I'm okay. But, oh, the eternal view of the parking lot, and the automated morning wakeup call, and the plastic cup in its plastic wrap, and the sign explaining that the environment is being saved by not washing your towels. Hilton should open a budget division called Purgatory.

So maybe some of the delight in seeing Kirk's new home was simply that it *was* a home. All summer and fall I'd kept in touch with the ongoing construction via phone calls, and if I got to Vermont for a day I'd stop by. He'd started in late spring by building the workshop/barn—twenty-six by seventy feet. "That was the most important job," he said. "As long as I got the barn finished I could always move into the double-wide," referring to the trailer the previous owner had left on the edge of the driveway. "But I needed the barn space to run the apiary."

He had to buy a tractor first, an orange Kubota, so he could cut the road and do other excavation. "I really enjoyed digging the conduit for the electric line," he said. "It only took a day and a half in the tractor. You understand why fossil fuel can be a good thing." With electric service in place at the barn site, he assembled his small crew: his godson, Heath; one of my former students named Corinne, who had no carpentry experience; her boyfriend, Nick, who had some; and a woman from Oregon named Kat, who had a Chihuahua named Anigo who guarded against visitors.

As the summer wore on, the frame went up, and then the walls and roof—it's a tall barn, with a full loft above the workshop floor, all laid out to make it easy to load and unload wooden hive boxes. By late summer everything was done but the shingles, and since it's well known that tar paper can get you through a single winter, Kirk had a moment to pause. Irene blew through at that point, but on the western side of the Green Mountains it was just a bad rainstorm. ("I didn't have my antenna up so I wasn't listening to the radio, and it was a few days before I found out what had happened," Kirk said. I marveled to think of the hours he hadn't wasted tracking the storm up the coast on the Internet, or staring at YouTube pictures of floating barns.) The end of summer was also the moment to collect the year's honey, but there wasn't much of a crop—even before the hurricane, the weather had been cool and damp. So there wasn't much money coming in.

"When I saw the state of the honey crop, I almost pulled the plug on doing the house the same year," he said. "I theoretically still had enough money in the bank, but I'm very cautious. I didn't want to risk using up my whole balance." (Needless to say, he was not borrowing to build.) "But I still had half my crop in storage from the year before"—stacks of fifty-gallon drums of honey—"and the people who had been buying my honey and reselling it in Massachusetts were doing so well that they wanted to buy the rest, and at a good price. All in all I had about $70,000 more than I thought I was going to. And that was enough." The crew, by this time more seasoned, picked up their hammers and returned to work to build the house. He had sketched out the plans on a piece of paper—an uninsulated workshop, a mudroom/sauna, and a single big room with Japanese doors that could partition off the bedroom if he wanted. "I knew I wanted to make the smallest house that two at the most could live in it without it becoming a mental health problem," he said. "And eventually I want a greenhouse attached to the south side. And I wanted it to be straightforward to build."

Which it was—the frame went up easily enough, but Kirk spent long hours finishing the inside. There's a beautiful built-in writing desk carved from a slab of redwood salvaged from the beer tanks at the original Rheingold Brewery in Brooklyn, New York. ("When I was sawing it up I could smell the beer," Kirk said.) The floor is made of pine boards, the same width and size as those used for the beehives; the oak trim is the same oak he uses for the stands that hold the hives. "It all came from Book Brothers sawmill in West Haven, where I've gotten the wood for my hives for years," he explained.

In some ways, then, he's living inside his own version of a hive, and it's just as free of chemicals. He bought the urethane to finish the floors from a plant in the small Vermont town of Hardwick; it's made from whey left over from dairies (the other half of the plant produces its own brand of native tofu). The house is highly insulated—I've never felt a draft. But when the windows are open the cross-ventilation cools things off immediately.

I ventured over the day after he'd spent his first night sleeping in the new home. "I was pretty exhausted," he said. "I'm not very good at transitions. I'm not a great traveler. But I love it here. It's a luxury suite compared to my old place."

Indeed. It was light, clean, quiet, airy, cozy. The few bookshelves were filled with his library: some Wendell Berry, some Chatwin, Tolstoy, E. B. White, Tolkien, *The Complete Book of Furniture Repair and Refinishing*, lots of old bee books, the journals of Lewis and Clark's expedition, a little Scott and Helen Nearing, and *Joy of Cooking*.

On the wall a small monitor tracks the amount of electricity coming off the solar panel on the roof, and the amount stored in the batteries—Kirk's house is entirely off the grid. "I've got six lights on and I'm using one-tenth of a kilowatt hour. The refrigerator in the workshop isn't plugged in, and it doesn't need to be because it's cold now, but when I plug it in it seems to go just fine." (When I checked back a few weeks later he said, "I've never seen the batteries lower than 75 percent, but if it's a problem I can just decide not to vacuum till the sun comes out.") Like Thoreau, he'd kept careful count of the total cost: "It took $75,000 to build the house, and $113,000 to build the shop," he said. "That includes hiring the backhoe, and what I paid for the tractor, and putting in all the road. If I didn't have to do the septic in the way the town insisted on, I could have saved $15,000. If I'd had more time I could have done more of the plumbing and electricity by myself, and that would have been another $10,000 saved right there." Still: lovely house, big barn, all for well under $200,000; even with inflation a bit more than Thoreau, but he was planning to spend the rest of his life there, not just a couple of years.

We sat eating bread and honey, with honey in our tea, and what struck me most was the quiet. The pan of water on top of the woodstove hissed softly as it humidified the house. That was it. Calm.

"I'm finally allowing myself to be exhausted," Kirk said, and I nodded in agreement. "I tried to take one whole day off a week through this whole thing, but toward the end I just wanted to keep going. I had it down to a science as we were building, how to wring every last bit of progress out of the day. It was a massive change for me, to work that hard, and I enjoyed a lot of it—it was exciting. But now I can get back to going whatever way the weather goes. In nasty weather it's a three-minute walk through the field to the shop. And when the sun's out I can go to the beeyards."

It was one of those uncomfortable moments when you suddenly realize you're in the wrong place—that you're a rube from the sticks in a sophisticated city whose customs you don't quite understand.

About three weeks after the president announced the delay of the Keystone decision, the insider news organization Politico invited me to speak at its swank "Washington Year in Review" symposium. Politico's young reporters had provided balanced coverage of the pipeline fight all fall, so even though I'd spent barely three weeks in Washington all year (and the most memorable nights had been spent in Central Cell Block), I found myself traveling down from Vermont to share a stage as an expert with Representatives Ed Markey (D-Massachusetts) and Lee Terry (R-Nebraska).

I was a little nervous—the street outside the hotel was crammed with limos, and the green room was filled with congressional aides punching their BlackBerrys. It didn't help that Terry, a Republican and reliable friend of the fossil fuel industry, had recently introduced a bill to force the rapid approval of the Keystone pipeline, overriding the president. But in Washington deep conflict doesn't necessarily seem to amount to much. Our panel was perfectly amiable—Terry and Markey were joking with each other like, well, colleagues, and I did my best to fit in, pointing out with as much gentleness as I could muster why the jobs figures Terry kept citing were bogus, I mean ridiculous, I mean overstated. It was all "agree-to-disagree" harmony.

Until, in passing, I said something that seemed so obvious it didn't even occur to me that anyone would object. In response to a question from a reporter about why Obama's delay announcement hadn't stopped the process, I said that clearly Big Oil wanted the pipeline revived, and that the industry was using the congressmen it funded heavily to make it happen.

Beside me I could feel Terry bristle. Are you saying, he quickly interjected, that we're "bought off"? And I suddenly felt bad, as if I indeed had said something wrong. I've taught Sunday school; I was raised to be nice. Had I hurt his feelings? My face reddened, I stammered, and I tried to say I didn't know anything about him in particular, that I was sure he'd eventually be part of the solution, and so on. But the frost stayed in the air, and I could barely focus on the rest of the questions. Was it really possible that people in Washington didn't understand what the rest of the country—left, right, and center—believes about them? That they take campaign money from corporations in return for doing their bidding? I went home and looked up Lee Terry in the database of "Dirty Energy Money" compiled by Oil Change International. Koch Industries had given him $15,500—they have a "direct and substantial interest" in the pipeline. Exxon-Mobil had given him $25,000. The Petroleum Marketers Association of America had tossed in $12,500. Conoco, Chevron, BP—all in all since 1999 he'd gotten $365,798 from the fossil fuel industry, and in the latest tally he'd voted with them exactly 100 percent of the time.

Sitting there on the dais, I was feeling rude, but also stunned. You really think this is okay? *To take money from people whose interests you'll then pretend to judge impartially?* For most of us, it's no different from going to a football game where one of the teams is paying the referees. But Washington has its own set of rules, and they're even more lax in the aftermath of the *Citizens United* decision letting corporations spend at will. One recent poll had shown that only 9 percent of Americans approved of Congress (compared with 11 percent who thought polygamy is a good idea, and 16 percent who approved of the BP oil spill). But it was their rules under which we'd fight the pipeline as the winter wore on.

Perhaps as an antidote to thinking about Congress, I lugged bee books with me all across the country, learning a steady stream of trivia about what may be the most remarkable animals on the planet. Some of the books were magisterial. In E. O. Wilson's recent and controversial *Social Conquest of Earth*, he explains that the bee that stings you when you disturb the hive "is a product of the mother queen's genome. The defending worker is part of the queen's phenotype, as teeth and fingers are part of your own phenotype." Okay, whatever. But there's also Simon Buxton's *The Shamanic Way of the Bee*, which describes his initiation into an ancient cult called the Path of Pollen. The process began with him sleeping inside a six-sided wicker basket and ended with him covered in a particular kind of honey that caused him to get an erection that "to my vision began to extend itself to an absurd length, some two feet in reach." (If you were trying to choose between the two books, Buxton's got a rave review from "singer/songwriter/pianist Tori Amos," and Wilson's didn't.)

Anyway, I know a lot of things: there's a cave painting from 15000 BC showing people taking honey from a wild hive; bees have two tiny hooks on each foot that stick like burrs, which is a good thing since the ones at the top of a swarm are, in proportionate terms, the equivalent of "a man, hanging by his knees and trying to support . . . thousands of pounds." Also, when the queen's time is up, the hive executes her in a formalized death ritual known as "balling the queen," which involves many bees forming "a solid ball around their mother, pressing harder and harder" till she dies. Honeybees pollinate a hundred thousand different plants. Their visual range spans blue, yellow, green, and ultraviolet. Nurse bees look in on each brood cell three hundred times in the course of a day. If human infants grew as fast as baby bees, your kid would weigh four tons within a week. A bee's wings beat 160 to 220 times per minute, "producing the note C sharp below middle C," and allowing it to travel as far afield as nine miles in search of food. If the hive gets too hot in summer, workers rush back with drops of water, and "all bees in the hive cease whatever they are doing to beat their wings vigorously, in unison." The resulting draft evaporates the water, resulting in a primitive form of air-conditioning. A single bee would have to fly fifty thousand miles to gather enough nectar for a pound of honey.

I can tell you about bees in Greece (Plato thought there were too many hives on the mountains of Attica, and Solon, the prototypical legislator, passed a law mandating a distance of three hundred feet between one beeyard and the next) and in England (soon after Elizabeth I perished, her royal beekeeper published a book called *The Feminine Monarchie; or, The Historie of Bees*, celebrating the general advisability of having women in charge). The hive that swarms when it becomes too crowded was held up to encourage Englishmen to leave their overpacked native isle for the new world: John Cotton, for instance, pointed out that "when the hive of the Common-wealth is so full, that Tradesmen cannot live one by another, but eate up one another, in this case it is lawfull to remove."

One place they removed to, of course, was North America, and Tammy Horn's excellent *Bees in America* demonstrates that by 1621 the Jamestown colony was importing beehives. In 1623 a new arrival planted apple trees on what is now Boston's Beacon Hill; when they fared poorly, he sent home for some bees to pollinate them. "Honey-hunting" for wild colonies "became one of the first American past-times," and it wasn't long before the hive metaphors were as widespread as the hives: colonial Americans, for instance, loved having "quilting bees," "husking bees," and the like. George Washington kept bees, and Martha Washington liked rose-flavored honey, which required a half cup of rose petals for every cup of the sweet stuff. The Continental Congress put a beehive on its first currency. Both Shakers and Mormons took bees as icons.

Bees went to war—or at least, by the time World War II rolled around beekeepers were urged to increase their wax production to "waterproof canvas tents, belts, and metal casing of bullets. . . . Airplanes waxed smooth with beeswax saved thousands of gallons of valuable fuel." The Pentagon has used bees to track land mines in Afghanistan and truck bombs close to home, but so far has not emulated the Roman soldiers who would thrust beehives into the tunnels of their adversaries, launch them in clay pots over the walls of besieged cities, and lob them onto enemy ships, where "they would so unnerve opposing sailors that they would often jump overboard to escape."

Could I go on? Sure I could. Summer bees live about six weeks. For the first three they stay in the hive, working as carpenters, guards, and nurses. In middle age they go out to forage—they're good for about five hundred miles, or three weeks, of flying before their wings are too tattered to go on. Queens lay 1,500 eggs a day. Since it's hard to see what's going on in the hive, observers have often made mistakes. Aristotle believed bees had teeth; somewhat less understandably, he believed that while bees made honeycomb, they used it to catch the honey that "falls from the air, especially at the rising of stars and when the rainbow descends." (Pliny the Elder called honey the "saliva of the stars.") Virgil thought bees were born from the rotting carcasses of cattle, and so did Samson in the Bible. When Jesus rose from the tomb, the disciples gave him a piece of broiled fish and a honeycomb; when the Buddha broke his long fast, it was with a honeycomb that a monkey brought him in the forest of Kosamba.

One book stood out from the rest of the library, though—a new volume called *Honeybee Democracy*, by a friend of Kirk's named Thomas Seeley. An apple-cheeked professor of biology at Cornell, Seeley seemed to be widely acknowledged as the bee man of his generation. He'd already written one classic book, *The Wisdom of the Hive*, that explained how honeybees managed to efficiently gather and process nectar. *Honeybee Democracy* attempted to answer another puzzle: when bees swarm, how do they agree on a new home? His answer, gathered by ingenious experiments over two decades, was not just fascinating; it also bore directly on the political fights we were entrenched in, and on the vexed question for me of working close to home or away in the world.

In the spring or early summer, a successful beehive gets too crowded—it's time for some of the bees to depart and form a new hive, finding a home and filling it with honey so they can overwinter and begin the cycle again. (Beekeepers obviously don't want swarming—they work to divide colonies before it happens, so they get a new colony instead of watching it disappear into the woods.) When they're getting ready to swarm, each bee will swallow a drop or two of honey, increasing its body weight by 50 percent—that gives them the fuel to make it to their new home and settle in. At the right moment, the queen and ten thousand or more followers will take to the air and alight in a huge mass, which usually looks like a beard hanging from a branch. (A third of the old hive stays behind, raising a new queen and eventually restoring the remaining colony to strength.) So—you've got ten thousand bees hanging from a limb. Here's the problem: *how do they decide on a new home*?

The answer is an exercise in consensual democracy. The mass of bees sends out a few hundred scouts, who scatter in every direction, seeking out suitable cavities in trees or walls. Over the years Seeley established that a dream home for a new colony has a capacity of about forty liters—big enough that the honey supply can last the winter, small enough that it's not too expensive to heat via the metabolism of the bees. Such a home, about the size of a wastebasket, is best if it comes with an entrance hole about 2.5 centimeters in diameter. Scout bees spend a half hour or more assessing a possible site, walking back and forth to figure out, apparently, how big and how cozy it is. And then they fly back to the swarm, which is where things get interesting.

Scout bees are coming from every direction, each with its own potential home site to report on. When they arrive, they do a dance to describe the location of the potential home—by watching the angle, Seeley and his students can determine how far away the site is and in what direction. And over time, he discovered that the bees danced longer and harder depending on just how well the new hole fit the colony's needs: a forty-liter cavity and the bees were practically on *Soul Train*; if it was only fifteen liters in size then the dance was feeble by comparison. At first, the scout bees put a "sizable number of widely scattered alternatives 'on the table' for discussion," Seeley observed. But as the hours pass (this process usually takes a couple of days) a few leading contenders emerge. And eventually, almost all the scout bees are dancing on behalf of a single site. By running tests with hive boxes of various sizes on desert islands, Seeley was able to prove the colonies almost always came to the right choice—and once they'd done so they flew there en masse, escorting the queen (who'd played no part in the whole drama) to her new home. By painting a lot of little dots on bee thoraxes, Seeley and his colleagues even figured out how big a quorum you needed before the holdouts for other sites conceded and everyone decided to join in a consensus.

Seeley describes how he put the insights from the hive to use in meetings of the Cornell biology department (oh, to be a bee on the wall at the first meeting!), where professors now cast secret ballots and where, as department chair, he attempts to stay entirely neutral while new information comes in. "My colleagues are always good 'scout bees,'" he reports, "and most are as uninhibited as a dancing bee about sharing their knowledge." He also compares the hive process at length to something I've known all my life, the New England town meeting. In our small towns, we gather on the first Tuesday in March to figure out the town's business for the year to come. It's sometimes heated—for most of us, garbage collection has a more profound influence on our daily lives than, say, the defense budget—but with the obvious need to work together, the help of *Robert's Rules of Order*, and the advantage of workable scale, we make decisions and adjourn for cookies. (I would never miss a town meeting, if only because my neighbor Barry King makes this particular kind with her own maple cream.) It's essentially Athenian, and it's worked for several hundred years, which is not as long as bees have been swarming, but considerably longer than other arrangements.

Here is Seeley's commonsense summation of what bees have to teach us about decision making; interestingly, they are principles that dominate at town meetings: "First, we use the power of an open and fair competition of ideas, in the form of a frank debate, to integrate the information that is dispersed among the group's members. Second, we foster good communication within the debating group, recognizing that this is how valuable information that is uncovered by one member will quickly reach the other members. And third, we recognize that while it is important for a group's members to listen to what everyone else is saying, it is essential that they listen critically, form their own opinions about the options being discussed, and register their views independently" instead of slavishly following a leader.

In every detail, this pretty much describes the exact opposite of how politics work in Washington, and explains precisely why, for twenty years, our elected officials have done nothing to make our earth more secure. We can't swarm—we're stuck here on our home planet—and so good decision making is at least as key for us as it is for bees. But we don't have frank debate, or foster good communication—we have Fox News. It's hard to imagine a hive that would, say, go over a fiscal cliff of its own making, or consider minting a trillion-dollar coin to solve its financial woes—or, for that matter, approve a pipeline when its most informed scouts had come back with the information that it might mean "game over for the climate."

In my town of five hundred people, on the first Tuesday in March 2012, we figured out the road budget and a new roof for the school, and we still had time to vote on a resolution condemning the *Citizens United* decision and the baleful influence of money in politics. And then I went back to D.C. and the ongoing fight.

And what a fight. It's not a fight about reason in any way—just a fight about power. The night of the president's announcement the previous fall, a senator said to me, "You need to understand that Big Oil never loses in this town, not ever. Don't expect them to take this lightly."

He was right. For five months we'd been on the offensive, battling against the odds for an improbable victory—improbable enough that the fossil fuel industry had never bothered mustering its full force. Now we were going to have to defend that win, as small and temporary as it was, against the industry's massed power. And instead of fighting out in the open, on the White House lawn, the battle now moved to the halls of Congress. I'd been taught in high school civics that the legislative was the most populist branch of our government, and that the lower chamber in particular was the "people's house." In truth, it's almost completely closed, a murky place where after a few hours of hearings the leaders decreed a vote on reviving the pipeline. It passed 234–193. The 234, we quickly calculated, had taken $42 million from the fossil fuel industry.

When members of Congress bothered to make actual arguments for the pipeline, they were easy to knock down. "President Obama is destroying tens of thousands of American jobs and shipping American energy security to the Chinese. There's really just no other way to put it," said Speaker of the House John Boehner. But if you hadn't taken $823,475 from the fossil fuel industry over the course of your career, there were, in fact, plenty of other ways to put it.

Jobs first: pipeline proponents routinely claimed the pipeline would create tens of thousands or hundreds of thousands of jobs, basing their figures on one "study," paid for by TransCanada, that claimed that, among other things, dozens of dance teachers would move to the pipeline route and open academies. In fact, a few seconds' thought makes it clear that a pipeline is designed to kill jobs, not create them. Yes, building it would occupy a couple of thousand men and women for two years and at good wages, but as a Cornell University study showed and even TransCanada eventually admitted, once the pipeline was built it would take thirty-five people to maintain it. Republicans such as Boehner had voted down every attempt by the administration to create more jobs; crocodiles would have been ashamed at the tears they were now shedding for the unemployed. Energy jobs will come when we commit to a future of solar panels and tight houses—they'll require lots of people swinging hammers. No one ships their home to China to have it insulated.

As for "energy security," it did make a certain brute sense to say, "Let's take oil from stable North America, not the volatile Middle East." It made sense, at least, until you actually looked at where the Keystone pipeline ended—in a so-called free trade zone along the Gulf Coast, where it would be cheap and easy to ship it overseas. Which turns out to be exactly what the refiners had in mind—the contracts were with companies owned by, among others, Saudi Arabians, who made it clear that the oil would be turned to diesel and end up in Latin America and Europe. A few of our House allies actually called the bluff on this one, offering an amendment to approve the pipeline if the oil stayed in the United States. Big Oil instructed its harem to vote it down.

But in political debate, unlike academic debate, the actual facts matter not at all. Politicians continued to insist that there were tens of thousands, hundreds of thousands, millions of jobs at stake and that the key to defeating Iran's nuclear plans lay in constructing a pipeline to Canada's tar sands. As the winter wore on and fuel prices became a key Republican talking point during the presidential primaries, the candidates kept insisting that gas would cost $2.50 a gallon (Newt Gingrich) or $2.00 (Michele Bachmann) if only we'd hook up Keystone. As usual, actual analyses showed just the opposite—since the new pipeline would actually ease a glut of oil in the Midwest, TransCanada had explained to Canadian regulators in official filings that it would raise gas prices across fifteen states, by some estimates as much as fifteen cents a gallon. But pointing out that fact didn't slow the flood of talking points; as government expenditure reports would later show, TransCanada's main response to Obama's decision to delay was to hire dozens more lobbyists and run thousands more TV commercials. (Which made us, I suppose, job creators.)

We saved our victory, such as it was, more through Republican ineptitude than Democratic valor. First Boehner attached a provision to a payroll tax cut bill trying to put the president on the spot by giving him sixty days to review the pipeline. But since it was transparent blackmail over a crucial issue, and since TransCanada hadn't even announced a new route for the pipeline, the president had some cover for saying, "Backed to the wall, I'll reject Keystone outright."

It was semi-brave, really—the president said those words the week after the head of the American Petroleum Institute promised "huge political consequences" if the pipeline were rejected, a threat he had the resources to carry out. But it was only semi-brave, since Obama also said he'd "welcome" a new pipeline application later on, and further that he'd "expedite" the southern half of the pipeline, from Cushing, Oklahoma, to the Gulf of Mexico. That was a cynical promise, since this stretch of pipe didn't cross a border and so the federal government had little authority to block it, but the president was trying to provide himself some political cover at the expense of rural Texans whose land would soon be seized by eminent domain and wrecked. (Those rural Texans immediately began to organize, and their brave stand would be one of the most inspiring parts of this whole fight.)

After that, the action moved to the Senate, still in

Democratic control—and this time it was the filibuster rule, much abused by the Republicans, that saved the day. Everyone knew that Big Oil had fifty votes for the pipeline as "moderate Democrats" started defecting, in this case "moderate" meaning "deeply in debt to the carbon industry." Joe Manchin from West Virginia went first—he'd taken more money than anyone in his party from the fossil fuel industry (and had run a memorable campaign commercial in which he'd used his deer rifle to literally shoot a copy of climate legislation that he'd nailed to a tree). Knowing someone's party, in fact, was far less important than knowing how much cash they'd gotten from whom. After one vote I talked to Steve Kretzmann, who compiled the dirtyenergymoney.com database. "I've been looking at this stuff for years and it still shocks me how the thesis continues to hold up," he said. "There are always a few outliers, but by and large they really are bought and paid for. It really is that simple."

Long term, this is why we have to amend the Constitution, win public financing for campaigns, and do the other vital work of basic governmental reform. But short term, we were casting about for answers. One was to buy a few thousand referee shirts and plastic whistles; in the weeks before the Super Bowl we rallied on Capitol Hill and then in congressional districts around the country, calling penalties on congressmen for "democratic interference" and "too much money on the field." (Good fun to get on a subway car with a couple of hundred other referees en route to the Capitol!) Another was more old-fashioned—having sat in at the White House, and having surrounded it with people, we now decided to flood the Senate with messages. May Boeve, who had helped found 350.org as a twenty-two-year-old and was now its executive director, spent a couple of days in nonstop phone calls with the heads of other environmental groups, persuading them that it would be worthwhile—and fun!—to work together on the project: at noon on a Monday in mid-February, everyone started e-mailing their lists asking them to get in touch with their senators. We'd given ourselves twenty-four hours, and a goal of half a million phone calls and e-mails, which would be by far the most concentrated blitz on any issue that the environmental movement had conducted in a generation, so we had no idea if we could do it. I took the train down from Vermont and spent the day at our tiny new office in Brooklyn, doing what we do these days when we can't think of a better approach: tweeting.

11:59 a.m.: "Okay, we're on. The next 24 hours will be crucial in the Keystone fight. You've never been needed more."

12:01 p.m.: "Half a million e-mails is a lot. I don't know if we can do it. But we're sure as hell going to try."

12:22 p.m.: I think this is good news. We crashed 350.org for a few minutes so many people tried to sign up, but it's back up. So go to it."

1:29 p.m.: "I'm told we just passed 100k e-mails to the Senate, one hour in."

By now it wasn't just the Sierra Club and the National Wildlife Federation spreading the word, but groups such as CREDO and MoveOn. The numbers were starting to pile up, even as Republican senators chose that afternoon to file their bill to approve the pipeline. In midafternoon we passed 350,000 ("I'm no numerologist, but this is a big number for us here"), and we were growing increasingly confident. I put on my suit and tie and headed uptown to tape that evening's *Colbert Report*. I'd done the show twice before, and it's by far the scariest thing on TV: Stephen Colbert is the nation's great satirist precisely because you never can guess what he's going to say next—it's like playing Ping-Pong with someone who knows five hundred ways to spin. But this night I was buoyant—sitting in the greenroom, just before the taping began at seven, I got to tweet: "Um, I don't quite believe it. We just hit 500k e-mails in under six hours. Whaddya say we just keep going?" I parried Colbert's happy ribbing as best I could ("Did you come down from Vermont in some car powered by . . . self-righteousness?"), and at the end he said, "So, just to make sure that no one out there in Colbert Nation sends an e-mail by mistake, what's the Web site we should avoid again?"

"Three-fifty dot org," I said, and before long we'd passed the eight hundred thousand mark. The next morning young people from the dozens of groups that had been involved filled cardboard boxes with printouts of the messages and carried them, by the dozens, to the office of Harry Reid, the Senate majority leader. They were in neckties and dresses; it looked like a scene from one of those old Jimmy Stewart movies, where the Western Union boys are stacking telegrams in the Capitol rotunda. My favorite picture of the day, however, was taken surreptitiously with an iPhone as someone passed the reception desk of a Midwestern senator. It showed the day's "call sheet," listing the issues that constituents had phoned in to complain about that day. There was one call on Obama and contraceptives, one on payroll tax cuts, two on unemployment insurance, and three on "horse slaughter." And 115 on the Keystone pipeline. We'd done our job.

For the next few weeks, the political pros did their job. Groups such as the Sierra Club and the Natural Resources

Defense Council and the League of Conservation Voters are masters of the inside game: they have dedicated lobbyists who shuttled briefing books and PowerPoint presentations to wavering senators, convinced donors to call the candidates they'd funded, and set up local phone banks. The inside game hasn't worked very well for greens in recent years, mostly because there's been no outside game to go with it. But everyone had worked for the previous six months to build that muscle. The echoes of our protests were still strong enough, if barely. The climactic vote came in early March, on a day when presidential candidate Rick Santorum was explaining why global warming was no problem ("Tell that to a plant, how dangerous carbon dioxide is"), and the low-lying island nation of Kiribati was announcing that it had bought a swath of land in Fiji and was beginning to evacuate its 130,000 residents. I was on the road as usual but watching Twitter as the votes were counted.

Big Oil was certain it would win. So certain that, as the vote was under way, the American Petroleum Institute put out a press release announcing that a "bipartisan Senate majority approves building the Keystone pipeline" and congratulating the body on "taking a bold step." But it lost—by two votes. A few minutes later API sent out a "corrected release," reporting instead that the Senate had "tried to take a bold step."

We knew there was still every chance that Big Oil would triumph in the long run—as I told reporters yet again that night, there are no permanent victories for environmentalists. And we knew that stopping Keystone wasn't going to stop global warming. Still, there's pleasure in beating the bad guys, especially when the odds are against you.

I'd grown up in Lexington, Massachusetts, and spent my

summers in a tricorne hat, giving tours of the Battle Green (which may explain why I've never confused dissent with a lack of patriotism—those guys were, after all, revolutionaries). But today I was thinking of those great men who grew up a generation later and one town over, in the transcendental forests of Concord. Henry David Thoreau: "If one advances confidently in the direction of his dreams and endeavors to live the life which he had imagined, he will meet with a success unexpected in common hours." Or his usually graver neighbor Ralph Waldo Emerson, in a more pugnacious mood: "When a resolute young fellow steps up to the great bully, the world, and takes him boldly by the beard, he is often surprised to find it comes off in his hand, and that it was only tied on to scare away the timid adventurers." I was no resolute young fellow, but many of my colleagues were, and for their sake I hoped that day's small win was the start of something, not the end.

4

SIMPLE

March is usually about my favorite time of year—the days are getting a little longer, but the snow is still deep. It is cold at night but warming during the day—days are spent skiing the woods for a couple of hours in the morning, then helping the neighbors in the sugar bush all afternoon, hauling sap and watching the boil.

But the second warmest winter in Vermont history meant that there wasn't much to ski on: we'd joked all season long about the "'nor-inchers" that would coat the ground white and not much more. It's somehow worse with the Internet—now you can check the long-term weather forecast every few hours and confirm that there's still no snowstorm on the way. It's like opening the freezer to find that, sure enough, still no ice cream.

About the middle of March, though, "annoying" started to turn into "spooky." Jeff Masters is the most widely read meteorologist on the Web, and on March 14, 2012, he reported that "a highly unusual weeklong heat wave is building over much of the United States, and promises to bring the warmest temperatures ever seen to a large portion of the Midwest." The "exceptional heat will also be exceptionally long-lasting," he predicted, moving east as the month progressed.

I took the dog out for one last slushy ski, and then waxed the boards for storage and oiled the chain on my bike. But I was unprepared for precisely how much reality was about to break over my winter-white head in the next few days.

The phone rang that evening, and it was the White House—which always makes me a little nervous, just because it's the White House. It's not as if Obama was calling me; still, Jon Carson is the guy who ran his field campaign in 2008, and who had met with me at the Hay-Adams back in August. He works down the hall from the president. (Like I said, I'd watched *The West Wing*.) When I called him it usually meant I was about to cause him some trouble, and vice versa. This was a courtesy call to let me know the president had just announced his travel plans for the next week. He was going on an "energy tour." It would start at a solar farm in New Mexico and then it would move to Oklahoma, and—"tell me you're not going to Cushing," I interrupted.

"We're going to Cushing," he said, and I believe I said, "Shit." He knew why I was angry, of course—that's why he'd called to soften the blow. In the game we'd been playing back and forth since August, this was the first time the administration really slapped us, and it stung. Here's why: Cushing, Oklahoma, is a huge oil depot. Keystone was supposed to go right through there. And when the president had blocked the Canadian crossing, he'd promised to "expedite" the southern leg of the pipeline that ran from Cushing to the Gulf Coast. Since TransCanada didn't need his permission to build the pipe, he was like a rooster taking credit for the dawn, but taking credit is a finely honed Washington art, and now the president was heading to Cushing to pose for pictures with a stack of pipe behind him. The administration was going to rub our noses in the fact that even our temporary victory was far from complete.

On the one hand, I understood why they were doing it. Gasoline had broached the four-dollar barrier, and it was the only stick the Republicans had left to shake at Obama's reelection campaign. The Republicans had spent the past few weeks fighting birth control, which proved unpopular with those Americans who occasionally engage in sex without desiring offspring. Gay marriage wasn't moving the needle for them anymore, either, so they were trying to climb out of the fever swamps of the Tea Party with that most traditional of election year issues, high gas prices. They were playing it for all it was worth—Newt Gingrich had a little gas pump that he would put on the podium while he talked. Presidents can't actually control the price of oil, but since a large slice of the electorate can't seem to figure that out, Obama needed some symbolism. Cushing was the place to stump for his "all of the above" energy plan.

But it was the wrong thing to do—soon we'd be seeing photos of bulldozers wrecking farms to make way for one more pipe (and, blessedly, photos of brave people standing up to them). And it was frustrating and confusing to everyone who'd worked to block the pipeline. I had no idea quite how to react. When in doubt, tweet. "Solomon proposed splitting the baby. Obama always actually tries to do it," I wrote. I went to bed but didn't sleep much, knowing the next week would include a stomach-churning stretch of e-mails and conference calls with our friends in this fight, and knowing that it wouldn't matter much what we did because Obama was focused on the campaign.

As it turned out, though, the week was far more surreal than I would have guessed, because the heat wave that descended on the continent was not just a heat wave. Before it was over, weather historian Christopher Burt would be calling it "almost like science fiction." Even as it began, it had that feel. Jeff Masters, the weather blogger, still lived near Ann Arbor, where he'd grown up, and he offered this account the next day:

> As I stepped out of my front door into the pre-dawn darkness from my home near Ann Arbor, Michigan, yesterday morning, I braced myself for the cold shock of a mid-March morning. It didn't come. A warm, murky atmosphere, with temperatures in the upper fifties—30 degrees above normal—greeted me instead. Continuous flashes of heat lightning lit up the horizon, as the atmosphere crackled with the energy of distant thunderstorms. *Beware the Ides of March*, the air seemed to be saying. I looked up at the hazy stars above me, flashing in and out of sight as lightning lit up the sky, and thought, *this is not the atmosphere I grew up with*.

Indeed not. That evening, in fact, the earliest F-3 tornado ever recorded in Michigan wrecked a hundred homes. And that was just the start. Saturday was the hottest St. Patrick's Day in 141 years of records in Chicago, the third straight day of record heat that would turn into a record-breaking seven-day run. Consider the reaction of the National Weather Service, a just-the-facts operation that grinds on hour after hour, day after day. It has collected billions of records (I've seen the vast vaults where early handwritten weather reports from observers across the country are stored in endless rows of ledgers and files) on countless rainstorms, blizzards, and pleasant summer days. If anyone's seen it all, it's the NWS, which is why it was weird to see this quote from its Chicago spokesman: "There's extremes in weather, but seeing something like this is impressive and unprecedented." The bureau's official statement: "It's extraordinarily rare for climate locations with 100+ year long periods of records to break records day after day after day."

"Rare," "unprecedented"—these became the standard descriptions as the heat wave wore on, day after day. Bismarck, North Dakota, went to 81 degrees, 41 above its normal for the day. International Falls, Minnesota, "icebox of the nation," broke its old temperature record for the date by 22 degrees. Winner, South Dakota, hit 94—two days before the official end of winter. Ninety-four degrees in the Dakotas in the winter.

Not surprisingly, the heat came with the highest levels of atmospheric moisture ever recorded in many places. (As I've said, a key fact for the twenty-first century: warm air holds more water vapor than cold; on average the atmosphere is already 5 percent wetter.) Torrential flooding rains broke out along the southern boundary of the heat, in parts of Texas where it actually helped recovery from the previous year's devastating drought. Apples were blossoming in winter in Michigan; in Atlanta, apparently, *everything* was blossoming in the record heat, because pollen counts broke the all-time records by huge amounts, coating the city like "powdered sugar on a doughnut," according to the *New York Times*. (One local carwash ran pollen specials—six dollars for a hose and dry, though as the proprietor admitted, "people need another one as soon as they leave.") Rochester, Minnesota, home of the Mayo Clinic, made news one day when its *low* temperature for the day broke the old heat record; soon, that was a commonplace. The reptilian part of me was enjoying the heat—I stretched out on the deck in shorts as I fired off e-mails and sat in on conference calls. But it was one part savor to three parts dread—I knew enough to know just how wrong this was.

When you have 100 or 150 years of records, the chances of breaking one are slim. But if you do, you'll almost certainly break it by only a degree or two—that's how statistics works. Unless something changes. When Mark McGwire took steroids, he broke the old home run record by nine. But this was as if the steroids had taken steroids. The new numbers weren't just off the charts, they were off the wall the charts were tacked to. This was not the old planet. This was a new one, the "Eaarth" I'd described in my last grim book, where the atmosphere contained enough carbon to change everything. At some point midway through the heat wave, the "Extremes" section of the National Oceanic and Atmospheric Administration's Web site simply crashed: too many records were being set for it to keep up with. Before the month was over, 15,785 high temperature marks would be set, compared with 1,385 places that saw their readings hit new lows. Jeff Masters did the regression analysis—this was something like a once-in-4,779-years event. But the numbers didn't really do it justice. It was watching meteorologists react that really clued you in. The veteran Minneapolis forecaster Paul Douglas, after blogging an exhaustive rundown of essentially impossible temperatures, just gave up. "This is OFF THE SCALE WEIRD even for Minnesota," he wrote. Ditto just about every other place else east of the Rockies.

The only people who seemed not to notice were running for president. The Republicans had been in Illinois for the heart of the heat wave, but, questioned about the warmth, Rick Santorum said, "This isn't climate science, this is political science," as if nature was engaged in exactly the same kind of spin as presidential candidates. (Later in the week, tape of a speech emerged in which he called climate scientists "Pharisees," which is a severely not-nice word in his world.)

Meanwhile, Barack Obama (who had watched a dozen people faint in record heat at one of his rallies) was, as promised, heading toward Oklahoma. All the green groups were phoning their connections in the administration to make increasingly plaintive pleas: "Don't actually pose in front of the pipe—pose in front of the oil tanks instead." But he did stand in front of the pipe, and his remarks couldn't have been much worse.

"You have my word that we will keep drilling everywhere we can," he said, while protesters led by the Indigenous Environmental Network were kept in a "free speech pen" at a park six miles away. "Now, under my administration, America is producing more oil today than at any time in the last eight years. That's important to know. Over the last three years, I've directed my administration to open up millions of acres for gas and oil exploration across 23 different states. We're opening up more than 75 percent of our potential oil resources offshore. We've quadrupled the number of operating rigs to a record high. We've added enough new oil and gas pipeline to encircle the Earth and then some." When you've gone from "in my administration the rise of the oceans will begin to slow and the planet begin to heal" to "we've wrapped the earth in pipelines," you've gone a long way.

One hoped—one prayed—that the president didn't quite believe his own "all of the above" policy. Not choosing is the opposite of policy. (Imagine, say, an "all of the above" foreign policy, where all allies were treated equally and Great Britain and North Korea were pretty much the same.) In climate terms it made no sense at all: drilling everywhere you can and then putting up a solar panel is like drinking six martinis and then topping them off with a vitaminwater—you're still drunk, you just have your day's full allotment of C and D.

This didn't smell like policy, this smelled like fear. The president's enemies had already begun running millions of dollars' worth of ads against him focusing on energy; the Koch brothers, who own a tar sands refinery, had just announced that they were putting $200 million into the campaign. Political gravity was reasserting itself. We'd won our small victory in the fall, when the president desperately needed to consolidate his base. Now, after that base had spent the winter watching Mitt Romney and Rick Santorum, they were fully consolidated, and the voters who mattered were in the middle. For the moment we'd kept our small win, but we were under no illusions that the president was on our side. A politician's job is to be on his own side. Proof? The president went later that hot week to a fund-raiser in Atlanta, held at the filmmaker Tyler Perry's thirty-thousand-square-foot French provincial mansion along the Chattahoochee River. "It gets you a little nervous about what is happening to global temperatures. When it is seventy-five degrees in Chicago in the beginning of March, you start thinking," Obama said. "On the other hand," he quickly added, "I really have enjoyed the nice weather."

The ritual nature of political action—they say something, we say something back, they push, we push, constantly keeping just this side of an imaginary line—was grinding me down. It ran counter to every instinct of a writer, which is simply to say what's true. The only time I felt completely honest was when an Associated Press reporter called, working on a story about how Republican candidates were painting Obama as an environmental radical. "Let me assure you that he's the furthest thing from an extremist," I said. "Really, nothing could be further from the truth."

The past eight months had taught me several things, such as how to send people to jail and what the news cycle felt like up close. But mostly it had taught me that the political world was not at the center of this fight. Even our rare wins couldn't stand up to the full force of the fossil fuel industry. We were playing defense—inspired defense, but there's no way we could slow global warming one pipeline at a time. There were too many—and too many coal mines and oil wells and fracking rigs. To have a chance we were going to have to go on offense. We'd need to take on the fossil fuel industry directly. So the afternoon of the president's awful Oklahoma speech, in between tweets and radio interviews, my old friend Naomi Klein and I had the first in a series of conversations about how we planned to do exactly that—how we planned to challenge the underlying legitimacy of the whole coal and gas and oil machine. As absurd as it sounds, we thought we'd found a wedge of sorts: a lever we might stomp on to crack some things wide open. But since Naomi was seven months pregnant, and since I knew that presidential politics would drown out every other story for months to come, that would have to wait till November.

In the meantime, we had months of political skirmishing to get through, and months of hot weather. Thank heaven there were the beeyards for a refuge.

Of course the beeyards weren't fenced off from the rest of the planet, either. I could leave behind my cell phone when I went there, but not the record temperatures. Since it had suddenly turned so weirdly hot, the bees were out and flying weeks ahead of normal. Kirk was making the rounds, taking the insulating cardboard packing crates off the hives and counting how many colonies had survived the winter. "I'm going to put my veil on," he said after our first stop. "These guys are a little nastier today than I'd expect. Usually in the spring they're reluctant to sting, they need everyone they can to build the colony. But maybe they're a little freaked out by the weather, too."

In thirty years in the valley, the earliest Kirk had ever seen bees gather pollen was the last few days of March. "Red maple is always the first good crop we get, and it's usually between the tenth and fifteenth of April," he said. But on St. Patrick's Day (dressed in beekeeper white) we kept passing trees starting to come into bloom as we drove between the beeyards at the northern end of his empire. The pollen year usually follows a predictable pattern: red maple provides the food that gets the brood-rearing season under way. Three weeks after the first big day, new bees start to emerge from their cells. Then the willows and sugar maples tide the bees over till the dandelions bloom—the "big event of the spring." Those usually emerge at the end of April, and by the tenth of May, Kirk said, "they're always in full bloom. And no matter when they start, the dandelions always seem to end May twenty-fourth or twenty-fifth. The dandelion is where they get a lot of the energy that lets them grow into a colony large enough to take full advantage of the clovers and make a honey crop"—that is to say, it's the dandelion that feeds the colony, and then the various clovers that let them put away the surplus that eventually feeds you and me. In a normal year those clovers—and the purple vetch and bird's-foot trefoil and alfalfa—show up around June 15.

This year, however, who could predict? We were under way insanely early—everything was suddenly budding out at once. If the weather reverted to anything even resembling normal, there'd definitely be a hard frost between March and mid-May. What would happen then? And would the soil, left naked in the sun so early, dry to dust by the time it should be growing clover? (By March 23, the state of Vermont was issuing "red flag" warnings to watch out for brush and forest fires.)

For the moment, though, you could soak in the sun and hope. We ranged the valley, piling up the cardboard crates and looking at the hives. "I always feel like I'm reclaiming my apiary this time of year," said Kirk. We visited the Mitchell farm, where the sheep watched suspiciously as we smoked the hives, and the Resnicks', and the farm of the six Dutch brothers, stopping to eat our sandwiches on the banks of Otter Creek and enjoying, despite its portents, the heat. Even the sweat. It had been a mild winter, but in Vermont even a mild winter means you're a little hunched against the cold; I could feel myself uncurling. Our count, by days' end, showed 562 hives alive. "That's up in the decent range," said Kirk, consulting a calculator on the kitchen table. "That's about a 74.9 percent survival rate. It's really much better than I thought would be the case. It gives me lots more options. I'm not forced to stretch. I'll be able to sell a hundred hives anyway," which at $220 apiece would make for decent cash flow. "Maybe a hundred and fifty. And the honey I've got left in the shop is supposedly worth twenty thousand dollars. So all that will give me enough money to work with this year, to keep things going. And hopefully there will be a honey crop come summer."

The bees, unperturbed and adaptable, were doing their best that day to ensure there'd be honey soon. At one of the beeyards, near the hospital in Middlebury, I sat on the ground for a while next to one box, watching them return in a steady stream. Each had its twin saddlebags bulging with a dull yellow pollen—probably from ornamental silver maples in a nearby front yard. Watching them, you began to realize how it was all possible. Each clump of pollen was tiny—the size of a kernel of corn on one of those weird baby ears you get in Chinese restaurants. But the river of bees never stopped; you could almost feel it adding up.

Bees lead the animal world in cheap metaphor production, but there are times when despite all precautions you simply can't avoid them.

Early in April, about a week after the heat wave had ended,

I was working in the Shoreham beeyard with Kirk. He was evaluating colonies, figuring out which ones were weak enough that they'd need a dollop of sugar syrup to get them through to the real spring. "Down near the lake here there aren't as many sources of pollen," he said. "It's a little shaky till the dandelions bloom. Till then it's still possible for them to starve." So we were lifting the tops of boxes to see how many bees were in each colony and holding up frames to see if they were collecting pollen. We found queens laying eggs and plenty of cells already stuffed with the bizarre spring's early pollen. In fact, the spring was so far ahead of schedule that a few bees were already hatching, the first of the year. "See how fast she's moving?" he said, pointing to one new bee eating its way out of the cell where she'd matured. "It's chomp-chomp-chomp. That's so healthy. When the mite infestations were bad, it'd be chomp, rest, chomp, wait some more, chomp."

As we worked through the morning, the air around us was filling with bees, flushed from the hive by the gathering warmth. They seemed unbothered by us. "I think it's going to be a good day for gathering pollen," Kirk said. "They're getting calmer and calmer." Indeed, some early risers were already returning with the day's first cargo, their saddlebags filled not with the dull yellow silver maple pollen of the week before but with egg-yolk yellow grains. "That's red maple," Kirk said. "You can tell how far away the tree is by how much they carry," he said. "These are pretty light loads, which means they've had to come quite a ways. They can calculate exactly how much they should be carrying."

How they do that figuring is something of a mystery, as is their ability to navigate. These ladies had likely traveled a mile to the red maple trees, and then navigated back to their particular hive, one of eighty strewn across this small bee-yard. They'd coped with wind and with the sun winking in and out behind clouds; their GPS was wonderfully precise. Or—and here comes the first part of the metaphor—comically precise. "Watch this," said Kirk, and he turned one box ninety degrees, so that the small slit door on the bottom was about eight inches away from where it had been before. Instantly, bees began stacking up like airplanes above O'Hare on a rainy day—they were lining up at exactly the spot where the slit should have been, and when it wasn't there they had no way of finding it. All they could do was hover desperately; it was hard for both of us to watch, and after about thirty seconds we relented and turned the box back around. The machine resumed operations immediately.

This doesn't mean bees are stupid. But as individuals they are *simple*—capable of prodigious feats of navigation, strong enough to carry great loads all day every day, but simple in their approach. They do one thing and do it well.

I thought of those bees that night in a lecture hall at Middlebury. A freshman, Kate Hamilton, had reserved a hall to show a new documentary about the Koch brothers, and she asked me to speak afterward, since I appeared briefly in the film. It was not a subtle movie: the Kochs, it insisted with lurid graphics and sad vignettes, are destroying the planet, trying to resegregate schools, suppressing the voting rights of minorities, and warping democracy with massive donations. (All, from my experience, exactly correct.) In fact, that day the Koch brothers' front group, Americans for Prosperity, had announced a $3.6 million swing-state ad campaign targeting the Democrats for slowing down the Keystone pipeline. It fired the usual volley of complete untruths—the pipeline would create zillions of jobs, free America from dependence on countries where the leaders wear bathrobes, and so on. If you run an ad like this enough times (and $3.6 million buys a lot of times) people believe it; I'd been getting calls from panicked senatorial candidates saying they were losing big on the issue. I did what I could—I spent the afternoon writing op-eds carefully laying out the facts. But I was under no illusion that they would count as much as the endless ads.

So I was a little frustrated even before I watched the movie, and realizing that the Koch brothers were doing the same thing on a dozen other issues made me surlier still. When the movie was over and the house lights came up, another professor and I took the stage for a discussion. A student asked what we thought of the *Citizens United* decision, and whether corporations should have the same rights as people. My colleague answered first, saying that more speech was always better, that companies were, in fact, composed of people, and so on. I could have answered empirically—by that point it was clear that the 2012 election was already turning into a battle of billionaires.

But I was thinking of the bees still. And so I said something a little more philosophical. "The reason we shouldn't count corporations as humans isn't that they're bad, it's that they're *simple*," I began. "They do the thing they do with great power—if you need a car built or an oil well drilled, a corporation is an amazing tool. It can gather resources from great distances, carry them exactly where they're needed, and combine various skills to produce something of great value from crude raw materials." It can, that is, gather grains of pollen and produce honey. But being powerful is not the same as being complex. Human beings are complicated. We have instinctual desires and cravings that drive much of our behavior, just like bees. But those are tempered by strange and wonderful forces outside ourselves, such as art—the making of something of significance out of nothing. Or religion, which as far back at least as the Buddha has taught us to suspect some of our instincts and cravings.

We can remember our ancestors, and we can imagine our grandchildren, and so sometimes we act in odd and counter-instinctual ways. We may cry when we see a hungry person, and even empty our pockets to feed him; in extreme cases we may give our lives over to that kind of service. Or we sometimes vote for politicians who will raise our taxes and give the money to the poor. Or we go to jail because we worry about global warming. The precise glory of humans is that we're complicated, and those complications are what rein us in—what might still, say, keep us from deciding to tap the tar sands of Canada or cut down the rain forests of the Amazon.

A corporation, far more wonderful in its abilities to execute a plan than any of us individuals, is nonetheless uncomplicated. It doesn't care much about the past and can't think very far into the future. If it does, its shareholders will rebel. It's less like a person than like a bee, at least in this regard. Given the power of speech like a human, it won't use it to reflect, to check itself, or to think about the larger good. It will simply put this new power to work on its single-minded goal of amassing wealth, just as, the Koch brothers did, sublimely unconcerned that their tar sands investments were threatening the planet.

In other words, if your goal is to efficiently tap the tar

sands, you need a corporation. But to decide if tapping the tar sands is a good idea, you need to keep corporations out of it. Their relentless simplicity will combine with their wealth to overwhelm reason, science, love. If you want honey you need a hive of bees. But if you were trying to decide if making honey was a good idea, bees would be the last creatures to ask. You know what their answer is going to be. In fact, if you get in their way they'll be a little perplexed for a while, trying to find the door. And if you persist in getting in their way, they're eventually going to get mad and sting. That's just how it works.

If you're curious about what a week of campaigning looks like, it looks a little like this: Friday night, I arrived in San Jose, slightly groggy, and found a strip mall Japanese restaurant via Yelp. (Yelp is excuse enough for the Internet—instead of bad hotel food, I had some kind of noodles in spicy broth in a storefront I would never have found on my own.) Woke up bright and early, and then traveled to a nearby high school for the "Green Teen Summit." Since I get maybe ten speaking invitations a day, too much of every day is spent saying no, but it's hardest to say no to kids, and I'm glad I was able to fit this in; a trio of girls introduced me, alternating sentences from a script they'd clearly spent hours polishing, and I did my best.

Then it was into the passenger seat of a biology teacher's Prius, and on to Berkeley, where I'd promised to speak to the City College. It was a few blocks and a world away from the famous University of California campus where I've given big speeches many a time—this was in a basement auditorium, improbably filled on a Saturday afternoon. My hosts were the Global Studies Club; my audience almost all first-generation college students, most the children of immigrants. I could tell that they'd spent little time thinking about climate change, but when I showed them pictures from around the world of people joining 350.org events, they understood their connection. Most of the people we work with around the world are poor and black and brown and Asian and young, because that's what most of the world is made up of. These kids, I knew, would help.

Because I was already there, I agreed to speak to a group of activists who'd hired a hall a few blocks away. Berkeley activists are a breed apart—someone parked near the entrance was handing out a leaflet of small type accusing me of being an agent of the Obama administration. The introduction lasted a long time because the emcee was engaged in a full-throated attack on imperialism, corporatism, and some other things. "We are socialists," he shouted. I apologized in my opening remarks for being more of a Methodist myself, and managed to get through my talk, which was, after all, an account of the ways in which the biggest corporations on Earth are indeed undermining its most basic systems. I received polite applause—but the first question I took was about the Rockefellers. They're oil barons, and aren't you an agent of them? I patiently explained that yes, 350.org has taken money from the Rockefeller Family Fund, which is where some of the heirs, many generations removed from John D. and his derricks, have put their money to philanthropic use, funding everything from Planned Parenthood to Clean Air Watch to MomsRising to the Alliance for Justice to Citizens Against Voter Intimidation.

Okay, my interlocutor continued, but what about chemtrails—shouldn't 350.org be fighting them instead? Chemtrails are an ongoing conspiracy theory, something about how government planes are seeding the atmosphere with chemicals to control our minds or change the weather. Or something.

I know this town was the birthplace of the Left—I've seen the pictures of Mario Savio leading the Free Speech Movement in 1964, and most of the people in the audience seemed sane and supportive. But, yikes, there really were a few crazies. Look, I finally said, you don't need to search high and low for a conspiracy. It's right in front of you. The most powerful industry on Earth is using that power to make sure it can keep dumping its waste in the atmosphere for free. There's no secret conspiracy required, no unmarked airplanes. Regular old airplanes will do the trick, and cars, and furnaces.

My colleague Jamie Henn picked me up at the wheel of a rental Nissan the next morning—Jamie was one of the original seven young people that founded 350.org alongside me, and during its first semester of operation his work earned him a grade. A good one, because he was a crackerjack organizer even at twenty-one. When we'd divvied up the whole planet, Jamie took charge of East Asia. He'd somehow managed to coordinate three hundred demonstrations across China for our first big day of action, while handling Japan, Vietnam, Burma, Cambodia (huge banners were hung from the ruins of Angkor Wat) as well. He also coordinated our "press office" and spearheaded our "development office." He's twenty-eight now, a seasoned veteran and an old friend, and we happily tooled east, making plans and sharing movement gossip.

We'd taken this Sunday mostly off from speaking because we wanted to visit a friend—Tim DeChristopher, an activist from Utah sent to prison for two years for civil disobedience at an auction for oil and gas leases. His protest was inspired—on the spur of the moment he bid on and won several of the leases, letting the government assume he was an oilman. Alas, he lacked the millions required to actually pay; in a classic overreaction, the same Justice Department that charged no one with perpetrating our banking meltdown accused Tim of financial fraud and sent him off to the pen. FCI Herlong is in California, but only kind of. We drove the long, straight route up to the crest of the Sierras, and then all the way down the other side into Nevada. When we reached Reno, we turned left, and headed up a two-lane road that eventually reentered California in a dusty desert. Herlong was not much of a town, its chief attractions being an army depot where they used to dismantle nerve gas shells, and the prison. I'd written the authorities weeks earlier, and after filling out many forms I received permission to visit.

I was glad I'd come because it set my mind at ease a little. Tim looked good—he also looked enormous, working out being the main way to pass the time. He'd survived a stint in solitary and was back in the general population at the minimum security unit, where he could spend most of the day outdoors. His fellow inmates were mostly there on drug charges or for white-collar crimes, so it wasn't too scary, and he had a large pile of books he was working through. Mostly, though, he seemed to be thinking about the movement he'd helped build. His bold action had drawn in many new activists and pointed all of us in the direction of civil disobedience. Now he was wondering what the next steps would look like; he'd be out by the next winter at the latest and was already looking forward to it. I filled him in on the latest in the Keystone fight—the way it had gone from a huge public battle to the subterranean fight boiling away behind closed doors in Congress, and my frustration at not being able to answer the millions of dollars in ads. Truth be told, there were moments I felt nearly as impotent out in the open as he did behind bars. We had to somehow make climate change a visible fight or we'd lose almost every time to the fossil-industrial complex.

I took one last look around the prison, with a faint shudder at the not entirely impossible thought that I'd end up someplace like it myself someday, and gave Tim another hug, and then Jamie and I set off back down the desert, over the mountains, and most of the way back to the Bay, stopping short at the university town of Davis, where I'd agreed to give a big talk the next day. Our hosts took us to the best restaurant in town, and I tried not to think about what Tim was eating.

The next morning, bright and early, I talked to a packed lecture hall at the John Muir Institute of the Environment, five hundred residents of one of the most eco-friendly towns in the country. Davis is the bike capital of America, its streets laid out to make cycling easier than driving, and I saw helmets resting under half the chairs in the auditorium—but people seemed to understand the fundamental point: that we can't actually solve global warming one bike path at a time. Setting an example is vitally important, and we have a moral duty to live the right way, but I'd left my solar panels at home and gotten on the airplane because addition alone isn't going to work.

You can weatherize your house, and your brother-in-law may see it and decide to follow suit, and then maybe he'd buy a Prius and his neighbor would. . . . If we had a hundred years, that's how it should work, the slow graceful cultural evolution to a new world. But chemistry and physics aren't giving us a hundred years. So we'll have to work by multiplication, too, by changing the ground rules, putting a stiff price on carbon so we change much more quickly than is comfort able. That implies politics, which implies movements, and we needed the folks in the audience to take part: I told them about Connect the Dots day, three weeks away, when around the world people would rally to demonstrate the toll global warming was already taking. We'd be raising banners on thawing Sierra glaciers ("I'm Melting") and staging underwater demonstrations on the dying coral reefs of the Pacific, all in an effort to conjure a movement that could make political change.

After the speech it was back in the car and a trip to Sacramento, for a lunch with the editorial board at the local public radio station, which was doing remarkable environmental reporting, and then an on-air interview for an hour. The host was engaging and engaged, but there was no chance of saying something new—the trick is to say something for the hundredth time and have it sound fresh, to mean it as you say it. And I did. Back in the car, west again to the crest of the Sierras. Jamie was at the wheel, and I had my iPhone set up to provide a personal hotspot so I could answer the day's accumulating e-mails as he drove. More reporters, more speaking invitations, and before long we were over the pass again and at the mountain town of Truckee, where the Sierra Business Council had asked me to visit. They'd just come through the worst winter in many years, the ski business off by 60 or 70 percent as brown January turned into brown February. It had been just as dismaying as our Vermont winter, and so they were receptive to my message. But maybe I pushed too hard—the first question came from a man who said he sold million-dollar software systems. "I try never to have an argument," he said. "I try to figure out how to make what I'm selling seem good for the other side. That's what you guys should be trying to do: make the oil companies understand that they can make money selling the sun."

At some level, of course, he was right. The sensible way out of this mess is for Shell and BP and Peabody Coal to become true energy companies, not oil and coal companies, and to devote their expertise and their incredible capital flow to building the next energy system on the fly. But scientists and policy wonks have been making that suggestion for a generation and it hasn't gotten through—yes, BP had promised in 2000 to transform itself to "Beyond Petroleum," and it had unveiled a nifty new sunflower logo. But it never invested very much in renewables, and in 2011 it shut down or sold off its solar and wind divisions, returning to its "core business." And the reason is simple: the oil companies are making so much money now that they can't quit. Unless we win the political fight to put a price on carbon they'll just keep doing what they're doing; I mean, these were companies that had melted the Arctic and then decided to drill it for more oil.

I don't think I convinced my questioner, though—he just kept saying he hated the way politics was "always about fighting." I hated it, too, especially at the end of a long day.

But the next day was a treat. We drove once more down the west side of the Sierras and toward the foothill town of

Nevada City. This whole weeklong trip had begun because an old friend, the poet Gary Snyder, had asked me to come and talk to his community. Gary was one of the original Beats—he'd read a poem himself at City Lights Bookstore that night in October 1955 when Allen Ginsberg first recited "Howl," and he'd been the model for Japhy Ryder, the most intriguing character in Jack Kerouac's *Dharma Bums*. But he'd escaped the wreckage of that scene, and the craziest parts of the '60s, by moving to Japan to live in a Zen monastery, returning around 1970 to publish *Turtle Island*, which won the Pulitzer Prize. Around that time he'd moved to a ridge above Nevada City, working with friends to build a Zendo (a space for meditating) and above it a house so singularly charming it was easy to see how he'd spent four decades there. I'd visited once before, years ago, and I could remember every detail: the way each room opened onto the outside world, the pond full of honking frogs, the summer study, the cherry tree in full bloom above the room with the tatami-covered floor where I laid down my suitcase.

We drank tea and talked—about fellow writers we both loved, such as Wendell Berry and Terry Tempest Williams; about the woods east and west, about words and gods and hopes and fears. For an afternoon—and it was the greatest present he could possibly have given me—I felt like a writer again, the thing I most wanted to be and at least for the moment really couldn't. We walked the neighborhood—about fifty people live along the ridge and meditate at the Zendo, and there are also mountain lions and bears and bobcats, and Emmi, the excellent apricot standard poodle who accompanied us wherever we went.

Gary gave me another gift that afternoon, one he couldn't

have anticipated. We went to visit the local farmers, John Tecklin and Angie Tomey, who run Mountain Bounty Farm, a community-supported-agriculture farm, or CSA. (Angie even has a floral CSA—she showed us newly cut bunches of heirloom narcissus, their yellow centers like yolks.) And we met their crew of apprentice farmers, ruddy from a day of planting tomatoes. But as we were following John's truck out to the fields where they do most of their growing, he suddenly pulled over and waved for us to follow suit. We got out, and I could hear it first: the buzzing of a beeyard. There were 120 hives spread out around us, but better yet there were six or seven swarms hanging in the branches above, looking just like the pictures in Tom Seeley's book. I knew there was some serious democracy under way, but the swarms looked for all the world like sloths hanging from the trees, masses of ten thousand bees clinging to one another.

I was deeply happy to see the bees drooping from the branches. I hadn't realized, I think, how much they were getting into my blood. Just the buzzing was soothing; I wished I had my bee suit so I could go closer and check the color of the pollen from the manzanitas all around us. But we couldn't tarry long: Gary drove us into town for a sound check for the evening's talk at a converted foundry. Nevada City is an old mining center, and the main street manages to look historic without being quaint. Gary had promised a good restaurant, but not until we were walking in did he add the other surprise—his old friend Jerry Brown would be joining us.

Thirty years before I'd spent an afternoon with the California governor when he was running for president. I was a college journalist and he was in the thick of the New Hampshire primary, but that didn't stop him from spending an hour with me explaining how we'd soon be communicating via wristwatch phones. Governor Moonbeam, they called him then, but damned if there wasn't an iPhone in my pocket today. We'd talked once or twice by phone about environmental issues over the years, but this was much more fun, the perfect remedy for a day when I was despairing about politics. Clearly he's a good politician—you don't get reelected governor of the nation's most populous state for nothing—but he's also a curious human being. He reminds me of my home-state senator Bernie Sanders, another practical man who acknowledges the pressures of the moment but also understands that those pressures don't excuse risking the future. That's precisely the calculation most politicians—including, I fear, Barack Obama—get backward.

After dinner we walked together back to the venue, where people were actually scalping tickets outside, something I'm not sure I've ever seen at a talk of mine. I gave it my best, partly because the governor and his wife were in the front row, each of them scribbling notes like graduate students, and partly because Gary had introduced me, vouching for me to this throng of neighbors. You might think it's a waste to preach to the choir, but the truth is, you need to get the choir fired up, singing loudly, all out of the same hymnal. The choir is always there, but most of the time it's just humming in the background, or singing so many different tunes that no distinct harmony emerges.

And when I was done, the first question came from a ninety-five-year-old woman. It wasn't a question for me—it was a question for the overflow crowd. "What are we going to do to help on this Connect the Dots day?" she asked. "Who's going to help me organize it?" Lots of the audience recognized her, and they made her tell the story of the time now fifty years in the past when she'd joined the Freedom Riders on the trip south. "I was glad you made people wear dresses and ties to your protest," she told me. "I wore my white gloves and my hat and even my veil when I went down south. And it—well, it confused the policeman a little, and he talked to me, and asked me about California." It was the perfect cap to the day—the perfect reminder that movements can work, that if we care enough we really can make massive change. I gave her a hug and went back to the ridge to sleep on my tatami mat, the best sleep in weeks.

So, of course, the next morning, as soon as we were far enough down the hill for the cell phone to begin to work, the news came that the Republicans in the House, on the first day back from their recess, had yet again attached approval of Keystone to the transportation bill. They hadn't held exhaustive hearings or advanced new arguments; they'd just done as their paymasters had demanded. And this time they'd carried seventy Democrats, giving them a veto-proof majority. The action was back in the Senate, and my contact at the White House was calling with a grave tone—we needed, he said, to go back to work on key senators. Even Barbara Boxer, the liberal California Democrat, was wavering. Meanwhile, Naomi Klein called with the news that the Canadian government—dismayed by the public protest over the Northern Gateway pipeline to the Pacific—was planning to shut down the public comment process, effectively ending independent review. From the democratic honeybees and the Freedom Rides back to the petrostate—I wanted to tell Jamie to turn around and head back to the hills.

Instead we made our way back to San Francisco and a long-planned meeting with the staff at CREDO, one of the most effective allies we'd had all year long. An offshoot of Working Assets, the credit card company that donates a chunk of its profits to progressive causes, CREDO is a mix of MoveOn and a cell phone company. I don't completely understand the business model, but I do understand that all year long they'd fired out one powerful e-mail blast after another, finding hundreds of thousands of people to help with the fight. Their webmaster, Elijah Zarlin, had performed the same task for the Obama campaign in 2008, so chances are you've read his writing—but instead of taking the government job that normally comes with that kind of résumé he'd gone to jail with us the summer before. Indeed, half the people around the table that afternoon had been arrested. An anti-Keystone banner hung on the wall, suspended by handcuffs. So we talked and plotted—how were we going to keep Senator Boxer on our side, and more important, win or lose, what would we do next? They were smart people—before me, I think, they'd figured out we weren't going to beat global warming one pipeline at a time.

And from there we had time for one more quick stop on the way to the airport—the old warehouse down by the ship docks where our actions team was busy painting banners for Connect the Dots day. They were hard at work on the giant black banner destined for the High Sierras, the one with huge white letters that said "I'm Melting." Hugs all around, and then to the terminal. I boarded the plane knowing that by the time I landed the Senate might have acted and the Keystone battle might be lost, but I was at least a little confident that we were building our forces for the bigger war.

But we didn't lose, at least not yet. When I got off the plane in Portland, Maine, our crew was already hard at work pushing phone calls to Barbara Boxer's office—she got 1,800 over the next couple of days and then issued a statement pointing out that "the Senate has already said no to Keystone," and adding, "We would like to keep off anything controversial that has nothing to do with this bill." Not exactly ironclad, but you take what you can get.

So I spoke to the Earth Day festival at the University of Maine, then moved on to D.C. in time to speak at Earth Day on the National Mall—which, given a steady downpour, meant about forty people standing in ankle-deep mud in front of a vast stage. But there was also a speech at the National Cathedral, where I got to see my hero Wendell Berry win an award, and then a talk to the young Democrats at Georgetown (all of whom wore three-piece suits!), and a sunrise speech at the Martin Luther King Jr. Memorial and a sermon of sorts at the Orthodox cathedral, then another at Swarthmore College the same evening. The next day it was on to Boston University to preach at Marsh Chapel, and then to New York for a benefit for a land trust at an Upper East Side women's club (there were excellent photos of past member Eleanor Roosevelt); the next morning out to Kingsborough Community College in far Brooklyn, where students come from 142 countries, back by evening to Boston for an interfaith gathering, and then a hurried trip to Albany, where hundreds of students had gathered to fight fracking, and then . . . This is what organizing is. You talk to people and try to get them engaged. You tell them about what people are doing elsewhere, so they can glimpse what they could do. It is—wait for it—a kind of pollination.

A tiring kind, and eventually you need to go home for a week and do the laundry and play with your dog and pay the bills and check in with the real world. The first day I got back (after a short trip to Burlington to help dedicate a field of new solar panels at the headquarters of the Episcopal diocese), I found Kirk in the beeyards gathering queens. He'd open a hive and start pulling out the eight frames, one after another, looking for the slightly larger queen so he could pluck her out. "It gets easier by the time you've caught thousands," he said, as I stared helplessly at the writhing clump of bees on each frame.

"The secret is, when you're catching queens don't look at anything else. Don't look at the brood and how it's doing, or the color of the bees, or admire any of the other miraculous things about the hive. Just look for the queen, and if she's not there move on to the next frame." He plucked one out of the mass, just like that, holding it by the wings. "Sometimes it's just how she moves. She waddles more than the others do. But you can just spot them, until you get tired. The way fatigue works is, you start to see queens where they aren't, and you start looking longer and longer at each frame, thinking she must be there somewhere. But she almost never is, unless you saw her the first time."

As he talked, Kirk stuffed the monarch into a tiny wire cage, along with three workers who would feed her. He sealed the cage with a plug made of honey mixed with confectioners sugar—when he stuck her in a new hive, the bees in that colony would eat their way through the candy to free her, a process that would take enough hours for the whole hive to get used to the idea that they had a new leader.

The queen's important, of course, but replaceable. And so easily; it made me think of the way European monarchs would marry off their children across the continent, and suddenly Britain would find itself ruled by someone who spoke German.

And so with organizing. I suppose, in some sense, I'm the queen bee at 350.org, but my influence is by comparison minuscule; in fact, none of the thirty or so of us who work there really "organize" our campaigns. It's a great planetary hive, less an organization than a loose campaign designed to mesh with the Internet ethos of distributed action. This was clearer than usual that week, as we careened toward the Connect the Dots day of action on May 5, 2012. We'd seen the polls indicating that more Americans were worrying about climate change because they were making the connection to extreme weather, and we figured the same was true around the globe. We wanted people to see the emerging pattern.

As it turned out, our friends in the fossil fuel industry didn't want people to connect those dots, and so they did their best to throw up some smoke. Friday morning, a few hours before Saturday would dawn in the Pacific and our rallies would begin, a fossil fuel booster club called the Heartland Institute erected a billboard in downtown Chicago. It featured an enormous image of the mug shot of Ted Kaczynski, the "Unabomber," looking bleary-eyed and appropriately murderous. Next to it, in giant letters, were the words "I still believe in global warming. Do you?" Apparently, Kaczynski had mentioned climate change in a manifesto after his arrest years before, and thus, according to Heartland's press release, the logic went like this: "The most prominent advocates of global warming aren't scientists. They are Charles Manson, a mass murderer; Fidel Castro, a tyrant; and Ted Kaczynski, the Unabomber. Global warming alarmists include Osama bin Laden. . . . The leaders of the global warming movement have one thing in common: They are willing to use force and fraud to advance their fringe theory." Subtle it wasn't. (Nor logical—Hitler himself had probably believed in gravity.) But if you've got chemistry and physics working against you, what are you going to do?

It seemed to us, though, that far from throwing the intended wrench in the works, it set up our Connect the Dots day more or less perfectly. We quickly released a letter addressed to Joseph Bast, the president of the Heartland Institute:

Dear Mr. Bast,

Earlier today you and the companies that support you announced a set of billboards suggesting that serial killers were pretty much the only people who feared climate change.

I'd like to thank you for doing that. The billboards are ugly, but they convey with graphic intensity the desperation of those who have fought on the side of the fossil fuel companies for a quarter century. I know you'd like your opponents to be murderers and crazed fanatics—that would make your job easier. But as it happens, this weekend will see rallies in most countries of the planet, arranged by entirely ordinary people who have already felt the sting of climate change. You can watch the pictures at 350.org—we'll be blogging them as fast as we can. What you'll see are people of every race and creed, united in the hope that the floods and droughts we've already suffered will be enough to sway the hearts and minds of our leaders.

Given the frantic and reckless nature of those billboards, I think it's safe to conclude we're making headway fast.

Best,
Bill McKibben

P.S. Oh, and we'll be writing to your sponsors, too, along these lines: Dear State Farm: Thanks for sponsoring the Heartland Institute, and its billboards insisting that those who fight against global warming are mass murderers. We're always more inclined to do business with those who call us serial killers!

Almost as soon as I'd pressed the send button on that letter, the pictures indeed began arriving. The first were from the Marshall Islands in the Pacific, where the sun crosses the international date line. Our crew there was underwater, all in scuba gear, holding a giant banner above a dying reef: "Your Carbon Emissions Kills Our Coral." After that the images poured into our Flickr photo stream faster than I could post the best of them on the blog. An early picture arrived from Rajasthan in India, where the wells in four villages around the Ranthambore National Park tiger sanctuary have gone dry; women in saris, holding black umbrellas against the heat, circled the empty cement hole. The next came from a sailor who'd just set a world record for solo circumnavigation of the planet—a task made easier, as he pointed out, by the fact that he'd had no problem getting through the Arctic Northwest Passage, a feat that would have been a fantasy a decade before. A series of truly striking images arrived from southern Sindh in Pakistan, where the International Organization for Migration was still trying to cope with the millions left homeless by epic flooding in 2010 and 2011—bearded men, veiled women, and small kids standing in front of the sandy barrens left behind by rushing rivers. Parched tea estates in Assam; survivors of the forest fires that claimed 173 lives in the suburbs of Melbourne; a group on a dry lake bed in Garissa, Kenya, where one man's sign simply said "Drought Is Killing Me." A throng along the shores of the Dead Sea in Jordan used helium balloons to illustrate the twenty-six-meter drop in water levels since the 1960s, while a big crowd on the beach of Tel Aviv used placards to proclaim their solidarity with low-lying island nations the world over. Micronesia checked in, as did the Maldives, where people turned out even though a military coup had sidelined the island country's democracy just weeks before. On Cape Town's Table Mountain climbers rappelled down from the top to hang a giant red dot above the nearly sea-level flats where hundreds of thousands of poor people live in shanty towns. At the Cathedral of St. John the Divine in New York, the largest church in the hemisphere, the priests blessed hundreds of bikes; in Bangkok monks gathered forlornly around a Buddhist temple wrecked by the previous December's flooding, which did damage equivalent to 18 percent of Thailand's GDP.

I could go on all day, because this went on all day—much longer, really, since we were following the sun. In London, hundreds played huge games of Climate Twister, connecting the dots between disasters with their bodies; in Boston, a squad of people in improvised uniforms, the "Metro Civil Defense Corps," stood outside subway stations handing out new maps that showed which lines would have to be converted to ferries as the seas rose. Sometimes the notes that came with the pictures were as poignant as the images. Organizers in Kampala apologized for the poor turnout at their rally, but their excuse seemed sound: "Yesterday the floods were at the peoples knees and crossing to the other side of the road was 5000/ (2$) to be carried on the shoulders." Azerbaijan, Iraq, Adelaide. In Aspen, it was skiers who staged a downhill race on grass to mark their nearly snowless winter. Pictures arrived from indigenous communities high in the Andes. One showed a mother in one of those distinctive bowler hats, standing with her daughter, a brown mountain in the backdrop. "In the Andes, mountain snow caps are receding. There are less and less glaciers and snow, which is affecting the runoff. No water, no life." Not far away, Bolivians rallied at Chacaltaya, the glacier which used to be the continent's highest ski area. Now it's not a ski area—it's just a pile of rocks with a rope tow. Activists gathered outside the Bank of America in Charlotte, North Carolina, to demand that it stop financing coal mines, and in Bellingham, Washington, to protest plans for a giant coal port; a few miles away, just across the Canadian border, protesters stopped five of Warren Buffett's coal trains before they could unload their cargo on ships bound for China. Eventually the protesters were hauled away to jail—but meanwhile in Portland, Maine, their allies were rallying on the site of the proposed terminus for a tar sands pipeline.

I sat there transfixed at the computer, hour after hour. Samoa, Mauritius, Bogota, Oberlin. It was like eating salted peanuts—it was hard to stop hitting refresh, even for a minute, especially since I could begin to make patterns out of the confusion. Dozens of rallies were happening along the path of Hurricane Irene, for instance, which had brought devastation to regions accustomed to thinking they were immune. I hopped in the car and drove over the spine of the Green Mountains and up Route 100, the corridor of maximum devastation in Vermont; it took me about forty-five minutes to get to the little ski town of Waitsfield, where huge piles of gravel were still washed up along the Mad River. Six hundred people had gathered for the rally—they cheered Senator Sanders and sat patiently through my talk before we gathered in the field to make our giant dot. A drone helicopter about the size of a microwave hovered overhead, taking pictures—by the time I got home the best of them was up on the Web, a shot that managed to show not just our gathering but the wrecked farm fields on all sides. And there it was, our tragedy, up next to pictures from Burma, where floods had destroyed thousands of hectares of rice paddies; from Cairo, where the rising price of grain was one of many triggers for the Arab Spring; from La Paz and Entebbe and Harlem and Ho Chi Minh City, where people gathered trash to make a giant mural. From Tajikistan and then from Michigan, where the freak winter heat had forced the fruit trees to bloom, setting them up for a killer April frost.

Midway atoll in the Pacific checked in with a picture of volunteers trying to rescue Laysan albatross chicks from rising waters, while in Araranguá, Brazil, residents remembered the first South Atlantic hurricane in history and the damage it had done to unwary coastal towns. In California, people hung signs on freeway overpasses ("Wake up and smell the permafrost"), and in Sundsvall, Sweden, they gathered amid the forests devastated by a freak Christmas storm. We received photos of a giant nighttime bike ride in Sinaloa; a barbecue at the Palm and Pawn Pub in Wagga Wagga in Australia, which March flooding had left completely submerged; and an undersea rally in Mozambique, where dugongs and other sea life are threatened by an acidifying ocean. From the Gura River in Kenya came a picture of a busted dam, and a group of farmers with a sign saying "In the last few months I was almost dry and now I am full." Their note said both drought and flood had caused havoc—that people were leaving the region. Rwanda, Palm Springs, Serbia, Senegal. So beautiful. Late in the day a picture of the tar sands refinery in Fort McMurray arrived with an anonymous note:

> I am an Oil Sands worker and risked my job to take this picture. Myself, along with the majority of my co-workers are ready for a renewable energy revolution. We need to stand together to eliminate the corruption that exists in this industry, start taxing carbon emissions, and creating green jobs for a sustainable future. We do not work in this industry because we like supporting large oil companies; we simply have no other choice. We want jobs that provide long term economic, social and environmental sustainability for ourselves, our country and our planet.

Manaus on the Amazon, Guangzhou in China. One of the last pictures to trickle in came from the League of Women Voters in Montgomery, Alabama—a crew of black and white women in the city once torn apart by Dr. King's bus boycott, now united by the fact that pollen counts were off the charts with the weird weather, and hence their kids were having to deal with extra asthma attacks. A few minutes later another photo appeared, this one of a single woman standing by the water's edge in the African countryside. The caption read: "Here, it is in a village of Gatumba in a country of east Africa which is called Burundi. In this photo, we are in the presence of a young lady showing with the finger there where was her house before being taken by the violent rain. Now you see it yourself, there is only a swamp of water."

By then the foul image of Ted Kaczynski on that billboard was scrubbed from my mind. These hundreds of thousands of people who'd spend their weekend holding rallies and planting trees didn't "believe" in global warming; it was the new reality of their world, making hard lives harder and shaking up the comfortable and the bucolic. And together we'd managed to put a human face on global warming. I figured we'd made our point when I checked the biggest of the climate denier Web sites the next morning, and there was a big headline about the day's events: "Connect the Dolts." Ah, but fighting dolts! And what do you know, State Farm had decided to stop backing the Heartland Institute—in fact, a whole passel of insurance companies had pulled out, not to mention Diageo, which imports Guinness. "Diageo vigorously opposes climate skepticism and our actions are proof of this," the company said. But I'm still going to stick to local beer.

Vermont's weather had been picture perfect for Connect the Dots day (though the best thing about global organizing is that you don't worry about the weather—you know that some places it will rain and some it will shine). But the weather held the next day, too, and so, though pictures were still pouring in, I got up early and drove down to the valley to join Kirk. Temperatures were rising through the fifties in midmorning, and the weather forecast said it would be near seventy by day's end—save for that weird and haunted March heat wave, this was some of the warmest weather of the year. But this time it was right on schedule.

Kirk was still rebalancing boxes of bees, trying to build the strongest colonies possible in the weeks before the clover would blossom and the honey crop would, with luck, be made. Sometimes that meant splitting apart colonies that had grown too fast—he was worried they'd start feeling so chipper they'd swarm, and he'd lose tens of thousands of employees before the flow even began. So he'd move a third of one hive off to a new beeyard and combine it with a weaker colony, and give it a new queen. "It kind of sets the clock back on these hives a little, gets them on my timetable rather than theirs. Or so I like to think," he said.

We were driving between beeyards on this high-spring morning, which meant we were crisscrossing New England's one real agricultural valley. Bordered by the Green Mountains to the east and Lake Champlain and the Adirondacks to the west, the Champlain Valley is a good-sized chunk of Wisconsin that somehow got stuck in northern New England. On a day like this, with the dust hanging in the air behind the pickup, it looks like a Ford commercial. But the soil's not quite good enough for true Midwestern scale: there are big dairies, and fields with a hundred acres of corn, and a few implement dealers with John Deere–green combines parked in neat ranks out front—but there are also plenty of grown-in woodlots where small farms failed generations ago, and fence rows tangled over with vines, and small orchards white with blossoms. Scale is so hard to get right. We were working in the shadow of a caved-in barn, where once some family had earned its living. Now the nearby land is used for a much bigger dairy that's going broke selling commodity milk, but also, down the road, a much smaller one, where half a dozen cows churn out ultra-high-end butter for a single restaurant. Mexican immigrants, most of them illegal, make the big dairies work; highly educated refugees from the city often run the artisanal cheese works and cideries. What's missing is the middle-sized operation, neither boutique nor big-box, the kind you think of when you think of farming. Kirk's apiary is pretty close to that sweet spot, though—a good year, remember, grosses him something like $50,000.

And he's done it not by piling on risk, but just the opposite: by building up resilience. There's plenty of possible trouble: mites, bears, cheap Chinese honey. You could try and resolve those problems with high-tech solutions, or by growing so big that you could ride out every bump. Or you could focus on durability—on the squat, hardy colonies now buzzing around us. In a way, Kirk was managing against failure as much as he was betting on great success; he was, in that sense, un-American. He was solvent and pretty much at peace, and doing something productive that didn't involve ever—ever—looking into a screen.

I fear that I haven't quite gotten that across. Kirk doesn't have a computer. Doesn't want a computer. Some of his friends set up a Web site, kirkwebster.com, that posts the essays he's written over the years for beekeeping journals, but he wrote them with a pencil and pad. He calls his customers on the phone—the old kind that connects to the wall. I'm not out to prove that this is a morally superior way of behaving. I spend most of the day on the computer, and the kind of organizing we do would be literally impossible without it; I think the Internet is one of the few wild cards we've got in the battle against corporate power. But, man, not having access to it saves a lot of time. Kirk doesn't ever have the task that takes up most of my day now, answering e-mail. His life doesn't seem Luddite or retro—it seems advanced. He's managed to choose the parts of modernity he needs (solar panels for the roof) and somehow kept the freedom to do without the parts he doesn't need. Bees aren't necessarily busy all the time—they spend the winter hanging out in a big warm ball—but they're pretty good about staying on task, about doing the things that actually need doing.

May 18, 2012.

It was exactly a year ago today that I first heard of the Keystone XL pipeline and began thinking about how we might bring it to wider attention.

Today, Mitt Romney unveiled the first TV ad of his general election campaign. Here's how it began:

VOICE-OVER: "What would a Romney presidency be like?"

VIDEO TEXT: "Day 1"

VOICE-OVER: "Day one, President Romney immediately approves the Keystone pipeline, creating thousands of jobs that Obama blocked."

Which is . . . weird to read. I don't know whether to be shocked at how (relatively) easy it turns out to be to make a stir, or dismayed that I didn't figure it out a long time ago. For better and for worse we managed to put something no one knew about at the center of the nation's political agenda.

5

. . . . .

ON THE ROAD AGAIN

Pandemonium! We were in one of Kirk's big beeyards on the Sunday of Memorial Day weekend, breaking down big hives into small colonies that would then be used for breeding new queens. The task involved lots of lifting and carrying, all of it conducted inside a roaring cloud of bees. "They're getting mixed in with bees from different colonies, they're getting taken to a new place, they've suddenly got no queen," said Kirk apologetically. I managed to get stung right through my bee suit—on this day we were dealing with wild animals, not domesticated creatures.

Which probably explains why we were telling wild animal stories as we worked. Pat Whitley was there in a bee veil and blue jeans (he was getting stung through the denim repeatedly); the beeyard was next to his land, by the house he built when he'd moved up from New Jersey some years ago. The standard rent is thirty pounds of honey. His son was the first to take a real interest. "I was watching from the house one day and saw a kid following Kirk in a bee veil," Pat recalled. "I thought, 'Oh, he must have a grandson or something helping him.' No, wait, that's James's shirt. He just got started into it, and I got into it to help him."

Pretty deep into it. Not long before, Pat had climbed halfway up a nearby pine to capture a swarm. A few weeks after that, a bear had crashed through this particular yard, knocking over a stack of boxes in the search for grubs and larvae. (Honey is an afterthought for bears, who really want the protein.) Pat went down to stack them back up, but "I didn't have all the right equipment, really. It took me an hour, and before I was done I had about fifty thousand stings. Well, twenty stings anyway. I was wearing wool pants, and Kirk told me later that bees don't like wool. Who knew?" Who indeed? As it turns out, wool retains some degree of animal odor even after you wash it a hundred times. So, wear cotton. Bees, by the way, don't like a lot of things, including dark colors—perhaps, authorities speculate, because it makes them think you might be a bear.

And bears really are trouble, which is why Pat—a dedicated hunter—decided he'd take down the bruin who'd gotten a taste for the local honey. "I was coming home from a function at the church one evening, and I could hear him crashing around," he said. "I snuck down, and I could see the shadow in the trees over there, but it was a dark night. If you can't see, you really shouldn't shoot. But I figured, so, he likes to come late. I got my blind set up—but the next night he was there by seven thirty, before I could get in place." After that, the bear had disappeared, and Pat thought he knew why. "On the school bus the other day, the kids were showing pictures of a bear. Apparently they've started feeding him marshmallows a couple of roads over, and he likes those better."

As our climate shifts, winters have gotten steadily shorter and people have gotten used to seeing bears out in the woods much later into the year. In fact, the state of Vermont had just announced it was changing the dates of bear-hunting season. "It used to end the first Wednesday of deer-hunting season," said Pat, meaning mid-November. "A decade ago there was usually snow then. But now bears can stay out a lot later in the year, so guys were seeing bears but they couldn't shoot them. They've extended the season, so that should get some more of them killed." The whole state was bear conscious at the moment, because our governor, Peter Shumlin, had woken up a few nights earlier to find a sow and three cubs dining from his bird feeders. Acting with the resolution one expects of a chief executive (albeit against the advice of all the wildlife experts, including the ones on his own payroll), he'd gone out on the deck to shoo them away, at which point they shooed him back in. "I was three feet away from getting 'arrrh,'" he explained to reporters, showing them cell phone photos he'd shot of the bears. What made the encounter memorable was a detail that the governor added: "Let's just put it this way, real Vermont boys don't wear pajamas. So the bear was better dressed than me." In some places that image might reduce gubernatorial popularity ratings, and indeed Shumlin's Republican opponent hired a man in a bear suit to follow the governor around for a few weeks—but most Vermonters seemed to think it was about what they would have done. It is, after all, the most rural state in the union—we've only got one real city, Burlington, whose population is forty-two thousand; Montpelier, where Shumlin works, has only about eight thousand residents (and is the only state capital in the country without a McDonald's). So we're used to seeing creatures.

The day before, for instance, the neighbors reported seeing a moose and calf in the meadow behind our house. That pleased me no end, not just because I like them above all other animals, but because they're in increasing trouble. Just that week Minnesota researchers reported that moose numbers had dropped by more than half in the past few years, according to new aerial surveys—the state estimated that as few as four thousand of the creatures were still roaming the north woods. And the culprit, not surprisingly, was climate change. Moose are exquisitely well adapted to the cold, which is to say that they're exquisitely badly adapted to the heat—above 20 degrees Fahrenheit they start looking for shade. (The need to cool off explains why they spend much of the summer standing in swamps and ponds.) By contrast, ticks love the new warm weather. In days of old, when I was in my forties, we had winters cold enough to kill them off—they weren't a problem up in the mountains. But no more—and so the Minnesota scientists were reporting that moose, who had evolved to deal with ten thousand ticks, now were carrying as many as seventy thousand at a time. The insects were driving them so crazy that they were scratching off their fur—biologists reported finding animals with only 10 percent of their hide intact. And then, what if we have the occasional old-school cold spell? "With no hair, if you're trying to survive in a cold climate, you're basically going to die from exposure," said one expert. "We may see little clusters hanging on in some areas, but it won't return to where we were before. Those days are gone."

As it happened, I'd found a tick in the shower that

morning, fully engorged with my blood. So I'd be spending the next week or so looking for signs of Lyme disease (flu-ish lethargy, a bull's-eye rash). Pat said his wife and son had both come down with Lyme the year before, fortunately recognizing it early enough that a course of antibiotics could knock it down. We never had Lyme disease in Vermont, not until the past few seasons. But already it was changing the feel of the place—there were people I knew who didn't really want to walk outdoors anymore. I could feel the reluctance myself a little—or at least the annoyance at having to tuck my pants into my socks and search my nooks and crannies when I came in from the woods. There are reasons enough for modern Americans to stay indoors, everything from obesity to screen dependency; all we need is another excuse for our ongoing denaturalization. It was a vicious cycle: warm weather breeds ticks, which causes people to care even less about the natural world.

But happily some folks remain engaged. We heard a gunshot from someplace nearby as we worked. "Another squirrel has succumbed to James Whitley," said Pat with a little paternal pride. "They steal the bird feeder food," he explained. The family raised chickens and bobwhite quail, along with their own hives of bees that Kirk had helped them start; they didn't have much use for predators. "We're an outdoor family," said Pat. "I like it out here in the beeyard. There's no radio or TV down here. It's just quiet time, a chance for me to talk with the kids as we work."

And the work for the morning was done. We had thirty-six new colonies, tied securely to the back of Kirk's truck. Up out of the valley we drove, into the mountains about twenty minutes away, where he kept his special isolated yards for raising queens. From the heat of the valley the temperature dropped ten, maybe fifteen degrees; a breeze was blowing in the sun-dappled grove of maple and white pine where we unloaded the hives. If all went according to plan, he'd be producing two hundred or so queens here every couple of weeks for the rest of the summer—since there weren't many flowering plants, no other commercial beekeepers kept hives in the neighborhood, which meant odds were good his untreated drones and queens would find one another, continuing to genetically improve his operation. The blackberry bushes were coming into lovely white bloom. "That will give these guys just enough to eat," said Kirk.

Since it was early June the grass was high all around Kirk's house, and since the leading edge of a rainstorm was blowing in from the west the grass swirled like the surface of a lake. Most of the farmers in the valley had already taken a first cutting of hay—the round bales were sitting in the fields, waiting to be fed to cows come winter. "We'll get it all cut at least once here, too," said Kirk. "And I'll bale a little bit of it—my godson is raising goats and they can use it. Most of it we'll just cut and leave, though. I know there are lots of farmers who view that with disdain—the grass is going to waste. But after just a single year you can see the fertility returning to this place. It gives you a real feel for just how much nutrition you take off a place when you feed all the grass to the cows, especially if you don't get the manure back on the soil."

I'd been pestering Kirk to tell me about the second great innovation in his beekeeping. A decade after he'd become the first apiarist in a generation to build a year-round queen-raising operation in the north, he shocked many in the profession in

2002 when he stopped treating his hives with chemicals—even organically approved treatments. And he did it in the face of the invasion of the dreaded varroa mites, which were killing off entire beeyards. I needed the story for reasons of my own, as you'll see—I was groping for ways to reorient our climate work.

"I'd always hated treating bees," he said. We were sitting at his table, driven inside by the rain, which we could hear pounding on the roof as we drank sweet tea. "But when the first mites—these were tracheal mites—arrived in the nineties, the predictions were so dire that I thought I had to use something or lose all my hives. So I used menthol, which was organic, and maybe not so objectionable, but even so it messed up the hives. The vapors were so strong that once it got a little warm it could just drive all the bees out of the hive, they wouldn't take care of the brood. So I stopped. And after a few years, I'd hear stories of other people not treating their bees, and they survived, too. I think if I hadn't had that experience I might not have had the courage to stop treating when the varroa mites came along—because the varroa mites were far, far worse." So bad, in fact, that their scientific name is *varroa destructor.* Tiny red specks, they feed on larvae, and in a few weeks they can destroy an entire hive.

"At first I treated my hives again, because people were saying that no colonies can survive otherwise," said Kirk. "And it was almost true, because the bees had no previous experience with this threat. It was like smallpox and the Native Americans. Actually, I was pretty sure some bees would survive—but I wasn't sure I'd survive as a beekeeper. The more I thought about it, though, the more I sensed it was risky either way. If I kept treating my hives, it would leave me dependent the way every other part of our farming economy has gotten dependent on chemicals. And that's a nightmare. I didn't want it happening to my favorite creature if there was any other way." He experimented with leaving a few colonies untreated. "As expected, they all perished." But he kept at it. He figured the small queen-rearing colonies, though the cornerstone of his system, were the least vulnerable to mite infestation because they weren't breeding continuously, which would starve the mites at certain points along the cycle. Beginning in 1998 he stopped treating those hives, and it more or less worked.

"But I wouldn't have been able to do it for the whole apiary without the Russian bees the USDA started importing," he says. "It was a rare piece of brilliant work that they did down there at the government bee lab in Baton Rouge." As the varroa mite started decimating the country's hives in the mid-1990s, a researcher named Tom Rinderer journeyed to the Russian Far East, along a rugged stretch of the Pacific called the Primorsky Territory. It had been settled a century before by, among others, Ukrainians, who'd brought their European honeybees with them to this new land near the Chinese border. The hives thrived, in part because the forests were filled with basswood trees, a great honey producer. But the area was also infested with the varroa mite, which had long coexisted with the Asian bee. "They crossed over, and when the beekeepers sent queens back to the rest of Russia that's how varroa originally broke out of its enclave and spread to Europe," Kirk explained. "But it also meant that this was the part of the world where European honeybees had had the longest time to work out some resistance to varroa—they'd lived together for a hundred years."

And so Rinderer brought back some Russian colonies, and

after they spent a couple of years in government quarantine, he was ready to release them to a select few breeders. Kirk bought two of the first available breeder queens the first year they were available—for $500 apiece, which ounce for ounce must make them among the most expensive animals ever traded. But it wasn't just the money he was risking. These were different bees, in some ways quite unlike the Italian strains that had dominated America's beeyards since the mid-nineteenth century. (Our continent had no indigenous honeybees—the only fossilized remains predate the last Ice Age.)

"The Italian bee has a whole series of characteristics that make it suitable for apiaries," says Kirk. "It's gentle, it's fertile, it builds up great big gigantic colonies. And they don't swarm nearly as much as other bees." They came from a warm country filled with plants that blossomed most of the year; winters were mild; life was good. "They almost mimic the stereotypes of the Italians," he said. Whereas the Russians—well, they're mimics, too. "They're more conservative," he said. "Once they've put up some honey in a sealed cell they have to be at death's door before they'll open up that cell and eat it." And they're used to a long hard winter.

They are, that is, a tiny bit more Kirk-like. Tough, frugal, endlessly hardworking. It was a good partnership. In April 2002 he stopped treating any of his hives and crossed his fingers. There were a few years of bad losses. He paged through his ledger books for me—"In 2004, I only managed to sell fifty-four hives; 2005, only fifty." But by 2006 things had stabilized. In fact, better than stabilized—there turned out to be a good market for untreated honey, and at a price twice that of the stuff from regular apiaries. And people line up to buy his untreated colonies.

Not, of course, that keeping chemicals out of his hives keeps them out of the bees. As news broke earlier this year that honeybee populations had crashed by half or more, attention turned to a class of pesticides called neonicotinoids that had been engineered into seeds and now appeared to be killing bees. Since, like teenagers, they roam the world, there's a limit to how much you can protect them. But at least they'd been well brought up.

"I still worry about varroa mite in the back of my head," Kirk said. "And we do have bigger losses. It's certainly harder to produce than it was before the mites. But we're doing fine. It's a combination of management—my system for northern beekeeping—and genetics, the steady improvement we get from our breeding program." And most of all it relies on a certain deep faith in the bees, on the idea that they know what they're doing. A faith that, pointed in the right direction and given half a chance, they'll figure out how to thrive.

I knew exactly why I was pestering Kirk about all this. All winter long, the thought kept nagging at me that we needed to shift our course at 350.org. The timing was odd, because people kept telling us how great we were doing—our global climate campaign was the largest thing the environmental movement had even seen; we'd built the first big green movement for the Internet age. The Keystone fight had demonstrated that we could rally people to go to jail. And at least for a little while we'd actually won something.

And yet I was wary. I'd watched how the big green groups had fallen into the trap of fighting the last war—their big Beltway operations were better suited for the 1970s, when they could lobby Congress with some hope of victory. And, more important, I was wary because we were losing. Badly. There were more carbon emissions and higher temperatures every year. The day I'd sat down with Kirk to talk bee genetics, the world's premier scientific journal, *Nature*, published a new paper authored by twenty-three high-profile biologists and climatologists, warning that we were on the edge of a planetary "state shift" that would leave the earth remarkably different than the one every human had heretofore known. "It really will be a new world, biologically, at that point," warned Anthony Barnosky, professor of integrative biology at the University of California, Berkeley, and the lead author of the study. "The data suggests that there will be a reduction in biodiversity and severe impacts on much of what we depend on to sustain our quality of life, including, for example, fisheries, agriculture, forest products and clean water. This could happen within just a few generations."

There was no way, fighting one lightbulb or pipeline at a time, that we could make a dent in that momentum. We had to figure out how to get to the source of the problem. Our strategy had been pretty much the same as everyone else's—go through the political system. Press the president, lobby the Congress, assemble at UN meetings. But what if we'd aimed wrong? What if the basic logic of our predicament meant we should be shifting focus?

I had an idea—that we needed instead to go straight at the fossil fuel industry. The basic problem we faced was that carbon carried no price—coal and gas and oil companies could pour it into the atmosphere for free, which undercut every effort at conservation or renewable energy. And the industry lobbied and donated and schemed endlessly to maintain that special break. No other business can put its trash out for free. That special privilege meant everything to the oil barons; it's why they were willing to spend huge amounts of money to maintain their position. (Huge in relation to what we could spend; in relation to their profits, they hardly noticed the campaign contributions and lobbying expenses. The return on investment for buying congressmen is truly remarkable.) So somehow we had to weaken that industry.

And I had an idea *how*, one that had been growing in the back of my mind for months, ever since that first phone call with Naomi Klein in March. She'd pointed me in the direction of a new study from a small group of UK environmentalists and financial analysts, one that contained three numbers that we thought might upend the stale climate debate.

Those three numbers—2, 565, and 2,795—seemed to offer a way to allow everyone to really understand the desperation of the climate debate, so I'm going to take some space to describe them. They are, I think, the most important numbers in the world.

The first of those numbers is 2 degrees Celsius, which is the only figure the world has ever agreed on about climate change. The only one. Do you remember the grand Copenhagen climate summit in the fall of 2009? If the movie had ended in Hollywood fashion, Copenhagen would have marked the culmination of the global fight to slow a changing climate. The world's nations had gathered in the December gloom of the Danish capital for what the leading climate economist, Britain's Sir Nicholas Stern, called the "most important gathering since the Second World War, given what is at stake." British prime minister Gordon Brown: "In every era there are only one or two moments when nations come together and reach agreements that make history, because they change the course of history. Copenhagen must be such a time." Danish diplomat Connie Hedegaard, who presided over the conference: "This is our chance. If we miss it, it could take years before we got a new and better one. If ever."

Of course, we missed it. Copenhagen failed spectacularly—neither China nor the United States was prepared to offer dramatic concessions, and so the conference drifted aimlessly for two weeks till world leaders jetted in for the final day. Amid considerable chaos (the State Department frantically calling the airport to find their Chinese counterparts), Barack Obama took the lead in drafting a face-saving "Copenhagen accord" that fooled very few. Its purely voluntary agreements committed no one to anything, and even if countries signaled their intentions to cut carbon emissions, there was no enforcement mechanism. Activists were angry (a Greenpeace spokesman: "Copenhagen is a crime scene tonight, with guilty men and women fleeing to the airport") and headline writers were brutal: "Copenhagen: The Munich of Our Times?" asked one.

The two-page voluntary accord did contain one scientific number, however—in fact, it contained it twice. In paragraph one it noted that "we shall, recognizing the scientific view that the increase in global temperature should be below 2 degrees Celsius . . . enhance our long-term cooperative action to combat climate change." And in paragraph two, it noted once more, "We agree that deep cuts in global emissions are required according to science, and as documented by the [Intergovernmental Panel on Climate Change] Fourth Assessment Report with a view to reduce global emissions so as to hold the increase in global temperature below 2 degrees Celsius." By insisting on 2 degrees, the Copenhagen accord ratified positions taken earlier in 2009 by the G-8, and the so-called Major Economies Forum. It was as conventional as conventional wisdom gets.

The number was first suggested, in fact, by a German panel in 1995, at a meeting chaired by Angela Merkel, who was then the German minister of the environment and is now the chancellor. It's not a hard and fast scientific line, of course. "There's no hard numbers to support 2 versus 2.2 or 1.8 degrees," said Josep Canadell, the executive director of the Global Carbon Project. So far, we've raised the temperature of the planet just 0.8 degrees, and that's caused far more damage than any scientist expected: half of the summer sea ice in the Arctic is gone, and the oceans are 30 percent more acidic. Given those impacts, many scientists have come to think that 2 degrees is, in fact, far too lenient a target. MIT's Kerry Emmanuel, the leading authority on hurricanes, wrote, "Any number much above 1 degree involves a gamble, and the odds become less and less favorable as the temperature goes up." Thomas Lovejoy, the World Bank's chief environmental adviser, put it like this: "If we're seeing what we're seeing today at 0.8 degrees Celsius, 2 degrees is simply too much." NASA scientist James Hansen, the planet's most prominent climatologist, was even blunter: "What the paleoclimate record tells us is that the dangerous level of global warming is less than what we thought a few years ago. The target that has been talked about in international negotiations for 2 degrees of warming is actually a prescription for long-term disaster."

If anything, environmentalists were even more dismayed by the target than the scientists. I was in Copenhagen as a volunteer campaigner lobbying various delegations, and I'd spent most of the two weeks wandering the vast conference center with a button in my lapel that read "1.5 to Stay Alive," a campaign mounted by the low-lying nations whose very existence was at risk. Dr. Albert Binger, the director of the Center for Environment and Development at the University of West Indies, emerged as a spokesman for the Alliance of Small Islands States: "Our ports, airports, roads and settlements will no longer be able to survive two degrees. Some countries will flat out disappear. You have a problem in the Pacific. Kiribati, Tuvalu, islands in Papua New Guinea and Fiji, across Asia and the Maldives." In Africa, beset by serial drought, some leaders urged even tougher targets. When Sudan's Lumumba Di-Aping, the chair of the G-77 group of developing countries, told African delegates at the conference that a two-degree rise in temperature was a "suicide pact" for Africa, many of them started chanting, "One degree, one Africa."

But environmentalists lost that fight. In the end, political realism bested scientific realism, and the world settled on the two-degree target—as I said, it's the only thing about climate change the world has settled on. By January 31, 2010, which was the deadline for signing on to the Copenhagen accord, 141 countries representing 87.24 percent of the world's carbon emissions had endorsed the two-degree target, and many more were added later. Only Sudan, Bolivia, Cuba, Nicaragua, and Venezuela have rejected it; the signatories include not just the United States and China, but also the rising powers India, Brazil, Russia, and Indonesia. Even the United Arab Emirates, which makes most of its money exporting oil and gas, signed on to the target. The official position of planet Earth at the moment is that we can't raise the temperature more than two degrees Celsius—it's become the bottomest of bottom lines. Two degrees.

The second of the three numbers is 565 gigatons. That's—again roughly—how much more carbon dioxide scientists say humans can pour into the atmosphere by midcentury and still have some reasonable hope of staying below two degrees. "Reasonable," in this case, means four chances in five, or somewhat worse odds than Russian roulette with a six-shooter.

This idea of a global "carbon budget" emerged about a decade ago, as scientists began to calculate how much oil, coal, and gas could still safely be burned. As I said, we've so far increased the earth's temperature about 0.8 degrees, which would mean that we're less than halfway to the target. But, in fact, most computer models calculate that even if we stopped increasing carbon dioxide now, the temperature would rise another 0.8 degrees—for the moment, that heat is being stored in the oceans, but it is working its way back into the atmosphere. That means we're actually more than three-quarters of the way to the two-degree target.

How good are these numbers? No one insists that they're exact, but few dispute that they're generally right. Tom Wigley, an Australian climatologist now at the University Corporation for Atmospheric Research, began making such calculations in 2001, with his colleague Sarah Raper. They looked at the futures envisioned by many of the giant computer simulations of the climate that have been painstakingly built at universities around the world over the past few decades, and tried to find where they converged. There are open questions, of course—precisely how sensitive is the climate to carbon emissions? How much carbon will the earth's oceans and forests soak up? These are hard to calculate, but as more data comes in the range of possibilities steadily narrows. The 565-gigaton figure actually comes from a 2009 paper written by one of Wigley's former postdoctoral students, Malte Meinshausen, who now works at the Potsdam Institute for Climate Impact Research in Germany. In the years since, universities around the world have continued to publish new simulations of the planet's climate. "Looking at them as they come in, they hardly differ at all," said Wigley. "There's maybe fifty models in the data set now, compared with twenty before. But the numbers are pretty much the same. We're just fine-tuning things. I don't think much has changed over the last decade." Bill Collins, who runs the climate science department at the University of California, Berkeley, agreed: "My personal gut sense is that the median has been robust for some time. . . . We're not getting any free lunch from additional understanding of the climate system." At current rates of carbon burning, we'd blow through that 565 gigatons in about fifteen years—before a baby born today makes it through high school.

But that's not the scariest number. The scary one is the third one—it comes from a team of financial analysts and environmentalists in London who calculated it in the summer of 2011 in an effort to educate investors about the possible downside risk to their stock portfolios.

The number is 2,795 gigatons—that's the amount of carbon already contained in the proven coal, oil, and gas reserves of the fossil fuel companies and countries (think Venezuela or Kuwait) that act like fossil fuel companies. It's the fossil fuel we're currently planning to burn—and the key point is that 2,795 is higher than 565. Five times higher.

James Leaton led the Carbon Tracker Initiative, which

used proprietary databases to figure out how much fuel each firm already had on the books. Oil and gas companies have to report their reserves annually to the Securities and Exchange Commission; reporting requirements are less strict for coal. ("I think the assumption is there's so much coal around why would you worry?" said Leaton.) But painstakingly, company by company, they built their list. The numbers aren't all absolutely perfect—they don't fully reflect the surge in the past two years in shale gas and other unconventional energy, for instance. And Middle Eastern countries have been known to play games with the reported levels of their reserves, in part to protect their quota shares under OPEC agreements. For the biggest companies, though, the figures are quite exact. Russia's Lukoil and Dallas's ExxonMobil lead the list of oil and gas companies, for instance. If you burned everything currently in their inventories, each would release just over 40 gigatons of carbon dioxide into the atmosphere.

Which is exactly why this is such a big deal. We've known the other two numbers for quite a while. The 2 degrees is the limit—call it equivalent to the 0.08 blood alcohol level below which you might get away with driving home. The 565 gigatons is how many drinks it takes to get there—say, the bottle of wine shared over the course of an evening that might let you stay just beneath that line. And the 2,795 gigatons? That's the five bottles of wine that the fossil fuel industry has on the table, uncorked and ready to pour. To repeat: *We already have five times as much oil and coal and gas on the books as any scientist thinks is safe to burn*. We'd have to keep 80 percent of those reserves locked away underground to avoid that fate. Before anyone knew those numbers, our fate had been likely; now, barring some massive intervention, it seems certain. It's a lot like nuclear overkill: we've got five times the carbon that we need to cook the planet already in our arsenal. But this time we're clearly planning to go ahead and push the button.

Yes, this coal and gas and oil is still physically in the soil. But it's already economically aboveground—it's figured into share prices, companies are borrowing money against it, nations are basing their budgets on the presumed returns from their patrimony. It explains why the big fossil fuel companies have fought so hard and so effectively to prevent the regulation of carbon dioxide—those reserves are their assets, the holdings that give their companies their value. It's why they've worked so hard these past years to figure out how to unlock the oil in Canada's tar sands, or how to drill miles beneath the sea, or to frack the Appalachians—the value of ExxonMobil is, more or less, the value of those reserves. If you told ExxonMobil that they couldn't pump out their reserves, the value of the company would plummet—a research report from the world's second largest bank, HSBC, showed that such a restriction would cut its stock price in half.

The financial analyst John Fullerton, who runs the Capital Institute, tried to put a number on it. At today's market values, he calculated, those 2,795 gigatons equal about $28 trillion. Which is to say, if you paid attention to the scientists and kept 80 percent of it underground, you'd be writing off more than $20 trillion worth of assets, much of it belonging to the richest people on earth. The numbers aren't exact, of course, but that carbon bubble makes the housing bubble look puny by comparison. Of course, it won't necessarily burst—we might well burn all that carbon, in which case investors will do fine. But if we do, the planet will crater. You can have a healthy fossil fuel balance sheet, or a relatively healthy planet, but now that we know the numbers it looks like you can't have both.

Do the math. 2,795 is roughly five times 565. That's how the story ends—unless we can leverage this new knowledge to somehow change the politics of this fight.

What Naomi had pointed out to me was that these numbers for the first time pinned down the fossil fuel industry—for the first time it was clear that they were planning to wreck the earth. Despite running hundreds of millions of dollars' worth of ads proclaiming their commitment to some kind of sustainable future, their business plans proclaimed in black and white that they were going to take us over the edge. If ExxonMobil burned its current reserves, it would use up 7 percent of the available atmospheric space between us and the risk of two degrees. Chevron would fill about 5 percent, and Shell about 4 percent. Just among those three, of the two hundred firms listed in the Carbon Tracker report, you'd use up more than an eighth of the remaining two-degree budget. This industry alone holds the power to change the physics and chemistry of our planet, and they're planning to use it.

Clearly they're cognizant of global warming—they employ some of the world's best scientists, and they're bidding, after all, on all those Arctic oil leases made possible by the staggering melt of northern ice. And yet they relentlessly search for more hydrocarbons—in early March 2012, ExxonMobile CEO Rex Tillerson told Wall Street analysts that the company would spend $37 billion a year through 2015 (slightly more than $100 million a day) searching for yet more oil and gas.

You could argue that this is simply in the nature of these companies. Having found a profitable vein, they're compelled to keep pursuing it. And so to call these companies enemies is to think of them as people, with some kind of free will. As I've said, they're more like automatons, or bees—admirably efficient because they're driven by pure profit.

But as the U.S. Supreme Court has decided, they are people of a sort, with a kind of free will. In fact, thanks to the size of its bankroll, the fossil fuel industry has far more moral agency than the rest of us—these companies don't simply exist in a world whose hungers they fulfill, they help create the boundaries of that world. Left to our own devices, citizens might indeed decide to regulate carbon and stop short of the brink—the most recent polling shows that nearly two-thirds of Americans would back an international agreement that cut carbon emissions 90 percent by 2050. Most of us would be perfectly happy with power from the sun or the wind. But we aren't left to our own devices.

The Koch brothers, for instance, have a combined wealth of $50 billion, meaning they trail only Bill Gates on the list of richest Americans. They've made most of their money in hydrocarbons, they know any system to regulate carbon would cut those profits, and they spent as much as $100 million on the 2012 elections. In the 2010 election cycle, the biggest player was the U.S. Chamber of Commerce, which took huge amounts of fossil fuel money and then outspent the Republican and Democratic National Committees combined; more than 90 percent of its support went to candidates who denied global warming. The week before the 2012 election Chevron made the largest contribution of the post–*Citizens United* era, spending millions to successfully ensure yet another Congress dominated by climate deniers.

• • •

So what should we do with this math? The plan Naomi and I had been turning over in our heads looked like this: the day after the presidential election we'd start a twenty-cities-in-twenty-nights road show around the country, beginning in Seattle. We'd gather as many people as we could each night, making sure that the bulk of them came from local colleges and universities. We'd go over the math, we'd have some music to charge people up, and then we'd send them off to see their trustees with this question: are you paying for our education by investments in an industry that guarantees we won't have a planet to make use of that learning?

If all worked out in the real world as it worked out in my mind, six months later the sparks we'd lighted would spread into a full-blown divestment movement of the kind that helped damage South Africa's apartheid regime during my years in college. There was already a nascent campaign, with students at Swarthmore College pushing especially hard on coal-mining stocks, and with groups like the Responsible Endowments Coalition and the Wallace Fund trying to spread the campaign. But I reckoned we might be able to take it up a notch by tying it more explicitly to carbon and turning it into a national campaign. Divestment wouldn't bankrupt the fossil fuel companies, but at least we'd alter the geometry of the political battle a little. I e-mailed my old friend Bob Massie—he helped lead, and then wrote a history of, the antiapartheid movement in America, and went on to help build the Investor Network on Climate Risk, so I figured he was as well placed as anyone to offer an opinion.

"The divestment movement allowed millions of Americans to cut through the obfuscation and express a clear and direct view that they did not want to profit from the destruction of the people of South Africa," he wrote back. "By supporting divestment, they rejected the halfhearted measures, feeble solutions, and institutional equivocation advanced by executives, institution leaders, and government officials. Given the severity of the climate crisis, a comparable demand that our institutions dump stock from companies that are destroying the planet would not only be appropriate but effective. The message is simple: We have had enough. We must sever the ties with those who profit from climate change—now."

Movements rarely have predictable outcomes. But any campaign that weakens the fossil fuel industry's political standing clearly increases the chances of retiring its special breaks. Consider Barack Obama's signal achievement in the climate fight: the large increase he won in mileage requirements for cars. Scientists, environmentalists, and engineers had advocated such policies for several decades, but Detroit was politically powerful enough to fend them off; only when it came under severe financial pressure did its resistance founder—and once the changes were made the industry began to recover, this time on sounder footing. A widespread understanding that the business model of the fossil fuel industry is—coldly and mathematically, not rhetorically—destructive to the planet might have similar effect. Or so we'd see. I spent much of the week outlining the plan to my colleagues at 350.org. They gulped, knowing at least as well as I did the amount of work this would entail. But they also knew as well as I did the basic fact that for all our work, we were still losing. And so they began to plan.

The odds seemed at least as good as that importing some bees from Siberia via the Ukraine might help a Vermont beekeeper survive the trouble that wiped out half the nation's hives. At any rate, it felt good to be in motion again.

Earlier I described a week of giving speeches around California; maybe I should describe another semi-typical week, this time involving some actual action. It began with nights in Boston (where we taped a public radio show before a huge live audience at the Paramount Theatre) and New York (public television, small audience, tiny studio). From there I flew to Columbus, Ohio, where a bunch of my cohorts from 350.org had been hard at work with anti-fracking activists from around the Buckeye State. Their organizing efforts became progressively easier after New Year's Eve, when the injection of high-pressure fracking water into a deep well managed to trigger a series of earthquakes. I'm not the kind of Christian who believes in signs from God—but you have to admit, earthquakes get your attention.

It dawned thundery, but this didn't seem to discourage the crowd. The wind did play a certain amount of havoc with the pair of helium balloons that were supposed to lift the banner high above the crowd—it eventually worked, but only because our stalwart actions chief Matt Leonard jury-rigged an incomprehensible number of extra guy wires and used four sturdy men in harnesses as anchors to keep it down. A few of us spoke our pieces, and then we marched the half mile to the state capitol, a line of at least eight hundred squeezing down to four abreast along a sidewalk. This was an impressive sight, but it was nothing compared to eight hundred people crowded into the ornate rotunda of the Ohio Statehouse, where the dome with the stained-glass state symbol rises maybe ten stories overhead. It didn't just look good—it sounded good, an echo chamber that made the standard chants into something dervish and grand. I think it was as close as Ohioans come to frenzy outside a football stadium; eventually, though, folks quieted enough to hear five or six locals give testimony to this "people's session" of the legislature. The week before, the actual legislators, beneficiaries of large lobbying donations, had passed a series of laws to make life easier for the oil and gas companies; our "resolution" didn't carry the same legal heft, but the outcry was a sign that the industry was facing a harder fight than they expected.

I had to leave before the rally ended, off to the airport for a flight to Chicago connecting to Munich connecting to Istanbul. The first leg went fine, and we were on the runway at O'Hare in line for takeoff, when word came that Air Force One was landing so the president could attend a wedding. I'd spent two weeks getting people arrested outside his home, so I was in no position to complain about the inconvenience, but the ninety minutes sitting on the tarmac meant I missed my connection in Germany and got to spend ten hours at the airport fighting for a seat on the next flight to Turkey. It wasn't entirely bad; it allowed me to get online to help with the Twitterstorm we were organizing to draw attention to fossil fuel subsidies as world leaders arrived in Rio for an environmental summit. All day long we were rounding up people to tweet with the hashtag #endfossilfuelsubsidies—we fell just short of our goal of setting a record (birthday greetings to Justin Bieber remain the top all-time tweet subject for a single day), but we were the number one trending topic around the globe. That is, the arcane and somewhat dull topic of fossil fuel subsidies drew more tweets than any subject on Earth that day. If you're going to sit in the Munich airport you might as well feel connected to something useful.

I finally talked my way onto a flight and got into Istanbul about eleven thirty in the evening, after thirty-six hours of travel, and without my suitcase—but there was a man with a sign with my name on it waiting for me. He spoke no English so I just climbed into his car, and we drove for an hour through the city till he deposited me at a dock—where another man, also speaking no English, loaded me aboard a small boat and disappeared into the wheelhouse, and we sped off down the Bosporus. I had no exact idea where we were heading, other than into the very choppy darkness, but I just sat there enjoying the thrum of the engine and the warm wind and trusting it would all work out. Sometime around two thirty a.m. we landed at the dock on the island of Halki, where an old friend, Father John Chryssavgis, was there to greet me and get me into bed.

I woke four or five hours later to the screech of seagulls and the travel-poster view of the sun-dappled Sea of Marmara, with its spray of islands, and the towers of Istanbul on the distant horizon. Halki is one of the few remaining seats of the Orthodox Church that once ran Constantinople; the nation's only Christian seminary was at the top of the hill above us, but the government of this overwhelmingly Islamic nation shut it down in the early 1970s. The Orthodox patriarch Bartholomew has fought to reopen the monastery, but as shepherd of the three hundred million Eastern Christians, he's also emerged over the past decade as one of the planet's most outspoken environmentalists. Indeed, in the fall of 2009, when 350.org was just seven college kids and me, he and the Dalai

Lama were two of the first world leaders to endorse a call for people to rally. "Global warming is a sin and 350 is an act of redemption," he said in a sermon—which made it easier to organize across the Balkans. It was one reason our first global day of action turned into such a wild success. So I didn't think twice when he asked if I'd join a few others for a small gathering on Halki.

I was a little groggy from long travel, but that disappeared when Jane Goodall rose to address the small assembly. I've heard her speak before, and I knew how she'd begin: with the greeting call of the chimp in the forest, a lovely soft hooting that reaches deep inside our primate selves. Before long it was my turn to speak—which is harder than usual when a patriarch for whom the proper form of address is "Your All-Holiness" is sitting in the front row in robes, long beard, and remarkable hat. I suddenly got a little nervous, but the day's subject, spirituality and climate, engages me deeply. I talked about the book I'd written years ago on Job, and about the clerics who'd come to Washington to get arrested, and about the basic blasphemy we're engaged in as we write the first chapter of Genesis backward—as we destroy the planet we were given. The patriarch speaks seven languages, Father John had told me, so I figured he could follow along, and everything was going fine until I mentioned the toll that climate change was taking on one of my favorite of God's creatures, those tick-ridden moose. Bartholomew instantly turned to one of the junior clerics, who turned to the next, who turned to the next, all searching for a translation of this strange word, which clearly described some kind of mythic beast. We finally settled on "horse with antlers" and continued.

That night Father John walked me up to the monastery

and seminary, officially closed by the Islamic government but still manned by a skeleton crew of three monks. We let ourselves into the church, lit only by a few candles in the twilight, and stood in front of one of the oldest and most beautiful icons, its base literally crumbled away by the hundreds of thousands of lips that had kissed it over the years. Most icons, he told me, were painted on wood, in part because the Orthodox had a strong sense of the sacredness of the material world. Outside, from the hilltop, we could see a rambling wood structure on the summit of the next island over, an old Orthodox orphanage that Bartholomew is trying to turn into an interfaith environmental center. God only knows if any of this will matter.

I was up before dawn, because I had to leave Turkey after only twenty-eight hours, and head, trailing a cloud of carbon behind me, to the global environmental summit in Rio. My companion for the hour-long boat ride back toward Istanbul was Jim Hansen, the NASA scientist who had really launched the global warming era with his 1988 testimony to Congress. I explained to him the logic behind the Do the Math campaign we were starting to plan, and he offered to help pull together a scientific advisory board—it was time, he said, for some other scientists to join him on the front lines of the fight. He's been arrested, testified in court cases, written innumerable op-eds, and taken more abuse than any of us from the climate deniers, all the while continuing to carry out groundbreaking research. As our boat skimmed the Bosporus, he told me about some of his latest work, including some preliminary modeling of the effects of what appears to be a rapidly accelerating melting of the great Greenland ice sheet.

"We have to look back into the paleo record to

understand what's really going on," he said, describing the remarkably violent storms that had marked a similar period in the Eemian interglacial period, 120,000 years ago. Researchers had found evidence of tempests that tossed debris far beyond the tide lines, the remnants of superstorms that he said could be one legacy of our rapid warming in the lives of his grandchildren—of whom he had pictures. (I countered with a day-old shot of Naomi Klein's brand-new baby boy, Toma.) "This is coming fast," he said. "Faster than any of our leaders understand."

Proof of that last point was on full display in Rio de Janeiro, where I arrived fourteen hours later to join the Rio+20 environmental summit. The first gathering, in 1992, had been a chaotic carnival, but a hopeful one; climate change was still a new problem—new enough that there was some reason to hope world leaders meant what they said when they promised to tackle it. (Though there was the ominous declaration from President George H. W. Bush on the plane that "the American way of life is not up for negotiation.")

At least the first President Bush had bothered to go. This time most world leaders stayed away, and those who did attend used the broadest platitudes. Hillary Clinton could think of only one American leader to quote, Steve Jobs. "Think Different." But nobody was thinking different, or even differently. They agreed on no targets or timetables for anything—theoretically they were for ending fossil fuel subsidies, but not at any particular point in the future. One sharp-eyed analyst went through the text, grandly titled "The Future We Want," and found ninety-nine uses of the weaselly formulation "governments should encourage" and fifty "governments shall support," but only three straightforward declarations that "we will" do anything at all.

Everyone in the sprawling conference hall seemed to sense the prevailing mood of futility—the celebrities were the same ones who'd been there twenty years before (Richard Branson, Ted Turner); the walls were papered with announcements of seminars so mind-numbingly dull ("Ecovision Turkey 2050"; "A Project for Human-Based Sustainability Through Ontopsychological Methodology") that it was hard to imagine actual human beings attending them. I'd been there about three hours, mostly talking with reporte[illegible]erate for something, anything, to cover, when a couple of young people asked if I'd join a protest they were planning. Of course—I seem to spend half my life asking people to join protests, and so am always eager to return the favor. Anyway, it sounded far more useful than any other thing I might spend the afternoon doing.

At the appointed hour we gathered outside the main plenary hall, and a leather-lunged young Canadian climate organizer, Cam Fenton, led us in ripping up our copies of the text—it was, he pointed out correctly, a sham. Then we sat down, maybe a hundred of us. I was the oldest (and over the next three hours would have reason to remind myself that sitting cross-legged on a stone floor gets somewhat harder as you age) and watched the young leaders conduct the session. Eight months' worth of Occupy assemblies had clearly taught them a good deal about the power, and the limits, of consensus.

Our moderator, a young British woman named Anna, quickly got the human mic up and running, and managed to head off a good-hearted suggestion that every comment be translated into Portuguese, instead assigning a few translators to the corners of the sit-in. She also constantly relayed messages from the UN security officials who were surrounding us in a phalanx of agitated consternation. They had declared this an "unsanctioned gathering," she said, and demanded we leave immediately or else "lose our accreditation." Which was, in its way, a serious threat—these young people had made their way from around the world, and the credentials around their necks were the token of their admission to the debate; some melted away into the surrounding crowd, unwilling to give up their legitimacy. But most seemed to understand that they were token in a different sense of the word, and decided to stay.

I clambered to my feet (a relief) to offer the suggestion that perhaps we should turn in our badges ourselves—that we could best demonstrate our disdain, our sense that the meeting had become a charade, by simply walking out and handing in our precious credentials. After some debate people agreed, and so our sit-in turned into a noisy, joyful march through the vast halls toward the exit. Lots of people joined in as word spread of what was going on. "The Future We Want Is Not Found Here," we chanted as we left. "Walkout, Not Sellout." Paula Collet, my 350 colleague who had done the most to organize our efforts at the conference, held a huge garland of everyone's lanyards and badges, which she presented to the security chief as we exited the hall. Our crew headed back to the crowded apartment we were sharing in time to catch the evening news—and found that we were the lead story across Brazil.

And when the proceedings mercifully ground to their official close the next day, we found we'd set the tone for the international verdict on the gathering. "Summit of Futility," said *Der Spiegel* in Germany. Juliet Eilperin, the *Washington Post* reporter covering the proceedings, put it like this: "The global environment summit concluding Friday, which drew nearly 100 world leaders and more than 45,000 other people to Rio de Janeiro and cost tens of millions of dollars, may produce one lasting legacy: Convincing people it's not worth holding global summits."

On the one hand, that's a shame—sooner or later, if we ever get serious about limiting carbon, we're going to need a global architecture that will make it possible. But at this point the international process was clearly going nowhere—and wasting effort better spent at getting to the root of the problem, the fossil fuel industry. Unless and until we weaken the industry's power, UN conferences are no more productive than Senate bills.

For me, that meant the most important parts of the gathering had been the chance to huddle with the leadership of the Natural Resources Defense Council (on the beach at Copacabana, with an army helicopter hovering overhead making us shout at top volume) and Greenpeace (over caipirinhas) to explain the idea for Do the Math. They got it, and they offered to help, and I left for the long flight home in a better mood than most of the delegates.

Man, it's a good thing climate change is a hoax, because by the time I got back to the States you'd sure have thought there was something to this global warming stuff. For a hoax it has excellent production values.

Consider: it wasn't just the 2,132 new high temperature

marks in June 2012. It was what went with them. Duluth, Minnesota, broke all its old rainfall records, and in an excellent cinematic touch, so much water flooded the city zoo that the seal escaped and swam down Grand Avenue. In the Gulf of Mexico, meanwhile, Tropical Storm Debby became the earliest fourth storm of the season ever recorded, and then dumped "unthinkable amounts of rain" on central Florida. (Giveaway movie moment: the nine-foot gator that washed into a Tampa swimming pool.)

Out west the largest fire in New Mexico history torched more than 170,000 acres, and then the most destructive blaze in the annals of Colorado burned on the edge of Fort Collins. But that was just the warm-up—it was Colorado Springs, in Waldo Canyon, where the nation really got to see what a wildfire looked like. Tweets and blog posts recounted the specific terrors: one resident wrote a harrowing account of driving his SUV across suburban soccer fields to escape the blaze, with "a vision of hell in my rearview mirror." It was cinematic in the extreme—the flames perfectly framed the famous chapel of the Air Force Academy on the very day the new cadets arrived. Another firestorm near the Boulder campus of the National Center for Atmospheric Research forced the evacuation of the planet's foremost climate scientists. I mean, c'mon.

The record heat moved east from Colorado, as records that dated back to the Dust Bowl fell with uncanny speed. Images of a farmer kicking the dust in his drought-ridden field—that old Hollywood staple—reappeared on the evening news; the scene worked so well that the price of corn and wheat shot through the roof. At 115 degrees, Hill City, Kansas, was the warmest spot on the continent, with farmers fainting in the field and workers burning themselves on tools left in the sun.

(The town is also home—cinematic touch—to a museum chronicling the history of the oil industry.)

An absurd number of catastrophes kept happening at the same time, just as in the best disaster flicks. On the last Friday of June, Washington set all-time heat records (one observer described it as like "being in a giant wet mouth of a dog, except six degrees warmer"), and then shortly after dinner a storm for the ages blew through—first there was five minutes of high wind, blowing dust and debris, followed by an explosive display of thunder and lightning that left four million people without power—which is to say without air-conditioning just as the temperatures got even hotter. By the first of July, the newspapers had just about run out of adjectives: stifling, oppressive, unbearable. The numbers were most telling. Atlanta, hottest temperature ever recorded. Ditto Raleigh. Ditto—the list was endless. In Goldsboro, North Carolina, the temperature was 103 and the humidity was 83 percent, which made for a heat index of 131 degrees.

None of this, of course, did anything to slow down the fossil fuel machine: the White House chose that week, in fact, to announce that the final permits for the southern half of the Keystone pipeline had been granted, that Shell could start drilling in the Arctic, and that it had auctioned off seven hundred million tons of Powder River Basin coal for the bargain price of $1.13 a ton. Nor did it humble the fossil fuel industry: Rex Tillerson of ExxonMobil told a New York audience, the same day that Colorado Springs burned, that global warming was "an engineering problem with engineering solutions." Right—an engineering problem. I could feel my desire to do some damage to the oil industry growing with every new weather report. We started signing contracts with booking agencies for the concert halls that would host our Do the Math tour.

But truth be told, it wasn't bad at all in Vermont. We were on the edge of the heat, so the temperatures nudged 90 once or twice—but mostly it was glorious high summer weather. Perfect for bees. Perfect.

The rest of the nation's agricultural system was buckling. Seriously buckling: the price of corn was spiking by the hour because across the Midwest the sun was withering the crop. The heat wave couldn't have come at a worse moment—as Bill Lapp, the president of an Omaha-based agriconsultant put it, "You only get one chance to pollinate over one quadrillion kernels. There's always some level of angst at this time of year, but it's significantly greater now and with good reason." In Kentucky, agronomists reported that "the corn crop is so desperate for water that kernels are aborting."

But the apiaries of Addison County were doing just fine. Or at least Kirk's apiary. Or at least the beeyard by the hospital, where we were at the moment.

Kirk smoked the first hive, lifted off the top, and gazed down. "That white? That's what we've been waiting for," he said. "That flash of white."

The white is the wax the bees use to cap the cells where they store honey. "They can't secrete that wax unless the honey's really flowing," he said. "Wow. If I can get all my colonies to look like this by the end of July . . . wow. This looks . . . prosperous."

It hadn't been a completely smooth morning. Earlier, at another yard, he'd had to burn a hive infested with American foulbrood, a disease that, before the invasion of the mites, was the most serious problem beekeepers faced. "In fact, around 1900 or so it seemed like just as big a problem as the mites are now," he said. Over time, though, bees seem to have evolved some resistance to the spore-forming bacteria, and Kirk thinks his bees, uncoddled by any treatment, are probably hardier than most. "I don't get it much anymore—I didn't see a single case till last year." Which is good, because foulbrood is about as nasty as its name. "You'll usually notice that the cappings to the cells where the bees should be hatching out have a little hole," he says. "The larva dies after it's sealed in the cell, and it melts down into brown goop, and the whole hive smells like rotten meat."

But it was just one hive, and he seemed unfazed, mostly because everyone else was starting to make honey. Most of the year, the bees are taking care of themselves—raising new bees, cleaning the hive, building comb. They're bringing in enough nectar to stay even, or using up stored honey, or relying on the sugar syrup Kirk provides when times are tight. But for a few weeks (sometimes just a few days, and some years not at all) flowers will open in such abundance across the valley that the hive turns into a honey factory. The bees bring in far more nectar than they need, and convert it to honey. If you keep adding storage space, they'll keep filling it up. Our job today was, in essence, to add attics to hives so that the bees had room to put up more and more honey. These attics are called "supers," and each can hold about thirty pounds of honey. In the very best days in the very best seasons on the shores of Lake Champlain, a colony can fill a super in two days—that is, they manage to produce fifteen pounds a day. "And this is nothing compared to, say, Saskatchewan," said Kirk. "I almost couldn't believe it when I saw it. The resource is so huge—the landscape is one-third canola, a third alfalfa, and a third wheat. Between the alfalfa and the canola there are unlimited flowers. They have hives on scales there that have added fifty pounds in a day."

Even fifteen pounds a day seems miraculous, though, given that the freight of pollen any one bee can carry is almost too light to weigh. "I love them as wild creatures," says Kirk, who was as relaxed as I've ever seen him. "It's just so amazing that they can produce much more honey than they need themselves, so I can take it without hurting them." And if that sounds a little like exploitation, well—in return he feeds the hives if they run short, and works year-round to see that they survive. He's out in the cold of November, wrapping insulation around hives, and in the mud of March making sure they've got plenty of syrup. It all works, at least in a year like this. Kirk's got a thousand supers, made by hand in his workshop during the winter. They're stacked six deep on some of the colonies. That means that if all goes right he could theoretically harvest thirty thousand pounds of honey. All hasn't gone right since 2005, mostly because of weather—"there just isn't any rule book anymore." But this has—so far—been a golden year. Kirk leads me down the country lane to see where at least some of the nectar is coming from. "That's a basswood tree," he said. "See those blossoms—it's just covered with flowers. A perfect honey tree." And the clover is showing up across the fields where the hay's been mowed.

So there's an easy moral to our story. On the one hand we have an agribusiness Midwest covered with uniform rows of corn and powered by fossil fuel—the same fossil fuel now driving the climate change that's causing the corn to wither in the impossible heat. It's a dead end. Here's Rex Tillerson again, the ExxonMobil CEO who truly doesn't know when to shut up. Expanding on his explanation that global warming is an "engineering problem with engineering solutions," he brightly added, "Changes to weather patterns that move crop production areas around—we'll adapt to that." Actually, we won't, not in the ways he's thinking. It's true that ExxonMobil has helped melt the tundra, but that doesn't mean you can just move Iowa north and start growing corn—the temperature may be cool enough, but there's no soil. And what do you do with the Midwest? Just let all that topsoil turn to desert?

No, the moral goes like this: we instead start producing a nation of careful, small-scale farmers such as Kirk Webster, who can adapt to the crazed new world with care and grace, and who don't do much more damage in the process—Kirk uses just enough fuel to get the pickup between beeyards. And that's almost kind of happening: in the past year, federal officials reported that after a century of decline there has been a net gain of thirty thousand farms in the country. And almost all the new ones are small.

But it's not going to be quite that easy. For one thing, the impossible weather can be at least as hard on good, small farmers as it is on large agribusiness concerns, precisely because they're small. Kirk weathered Hurricane Irene just fine, because the rain didn't fall that hard on this side of the mountains. But twenty miles to the east, great small farmers had their great small farms washed away. One young couple in Cuttingsville had built a lovely CSA farm, feeding dozens of subscribers, along the Mill River—which turned into a farm-wrecking torrent in minutes. As the local newspaper described the aftermath, "All that's left are thousands of rocks—permanent reminders of the land they once had."

There's another problem, too: raising farmers is just as hard as raising any other crop. Maybe harder. Kirk and I sat in the sunshine, eating lunch. "I think about it all the time—it's what I want to do more than anything else: have some apprentices that I can really teach to do this work," he said. "But I'm afraid I'm also somewhat jaded and cynical. Our culture has kind of abandoned farming." This is indisputably true—America has twice as many full-time prisoners as full-time farmers. Once farmers made up half the population; now it's under 1 percent. It's true that young people are starting to return to the soil, and often with fine technical skills gained at a growing network of college gardens or working as "WWOOFers" (from the volunteer-linking organization Worldwide Opportunities on Organic Farms) on organic farms. But the hard part, Kirk said, is the culture, not the agriculture.

"I think about the few farmers I know of who have really been successful over long periods of time: the Amish, the French market gardeners outside Paris. Both decided they had to remain a little separate from the rest of society. That's the huge challenge for all of us who grew up here, in regular society. How do we do this job and still have friends, families? Take time: scheduling five days a week is what we're used to, but it doesn't work that way with farming. You have to go with the schedule of nature and the weather. And the conception of money is completely different. Last year in this beeyard, there was nothing in those honey supers—no income at all because the weather was so bad. This year may be different—but you can't say how much money you're going to be making. I can't say, 'I make a salary and it's this much.'

"Where a lot of farmers fail is in trying to make farming fit into the time and economic parameters of the rest of society," he said. "I love the kids at the Middlebury college garden, but I don't know how many can make that transition." There's a better chance with people who already have a trade—like Pat Whitley, the homebuilder who'd been helping us in the beeyard near his house the week before. "They know sometimes you have to work when it's time even on Sunday," Kirk said. "You have to be skillful with your hands, and kind of an athlete—at the moment, given our society, it's only the most determined individuals who can really succeed at it." And even that success comes at a price—as I've said, I know for a fact that one sadness in Kirk's life is that he never found a wife willing to share his work. It can be hard.

But not today. Not in the sun, with the flash of white in every super that means honey. On a continent currently baking—a continent that feels like a giant never-ending disaster film—this is the closest thing to a sweet spot.

Ten days later I was in the beeyards again, with Kirk and with my fourteen-year-old niece, Ellie, who was staying with us while her mother underwent chemotherapy for breast cancer. Kirk's a gentle soul and Ellie a brave one—he helped her into a slightly oversized bee suit and veil, and within minutes she was carrying boxes through the small cloud of bees. She'd seen a moose earlier that day near the house, so it was a good day for creatures.

Kirk wasn't quite as high as he'd been in late June. It hadn't really rained across the Champlain Valley since, and the lushness was abating, replaced with a slightly dusty aspect, like a car that hasn't been washed in weeks. It wasn't a drought, by any means. The rest of the country was *definitely* in a drought, the worst since the 1930s. The USDA had just announced that a thousand counties in twenty-two states were in a state of disaster, the largest declaration in its history. A year after record rainfall had flooded the Mississippi so badly that engineers had to blow up levees to save cities, farmers were now finding twenty-eight-inch cracks in the soil. The *New York Times* reported that across the grain belt corn plants looked "like house plants better suited for the windowsill." Plant tassels were shedding pollen, but without water it wasn't fertilizing kernels. "You couldn't choreograph worse conditions for pollination," a University of Illinois agronomist explained.

That alarmist left-wing rag *Bloomberg Businessweek* ran a story under the headline "U.S. Corn Growers Farming in Hell as Midwest Heat Spreads." The price of corn and soybeans were setting new records daily, as a hoped-for bumper crop turned into a slow-motion disaster—farmers estimated that some fields were losing five bushels per acre per day of potential yield as the heat wore on. In hard-hit Missouri, where farmers were recalling the Dust Bowl years, one grower told the local paper, "Bumper crops are totally out of the question. Average crops are totally out of the question. Let's just go and see what we can salvage."

There was nothing like that in Vermont; by any standard

we were the luckiest corner of the country that summer. The epic heat wave had passed to the south, and we'd had temperate, beautiful days. But very little rain. The hives we were opening were more hit or miss than the ones we'd worked those few days before. A few were fat with honey, the glistening white combs overflowing the frames; we stuck more supers on top. As many were just holding on: to Kirk's eye, it looked as if they'd swarmed sometime in the last month, with two-thirds of the bees abandoning the hive and heading off with a queen into the wild. "I don't know if it's changing conditions or what, but it seems like bees swarm now at times when they wouldn't have dreamed of doing it before," he said. If he visited every beeyard every few days, he probably could prevent some of the swarming, but that would use up the time he needed for breeding queens. "Anyway, part of what I'm trying to breed is a bee that doesn't swarm too much, so it's a fine line," he said. "And it's also true that what's left of these colonies, starting over after they've swarmed, have a better chance of making it through the winter. They'll be concentrating from the start on overwintering. And for my system, having bees next spring is as valuable as having honey right now." Kirk was smiling; Ellie and I, driving off, agreed that, whatever the long-term meaning, we liked those frames dripping with honey.

Still, taking some short-term lumps for long-term gain was a good image to have in mind that night, when things turned a little dark at 350.org. The endless heat wave had been wearing on—Washington, D.C., had just passed its ninth straight day above 95 degrees, the longest in its history, but exactly the kind of stretch scientists say will become ever more common as the planet warms. Our American organizers had more than enough work battling fossil fuel subsidies, but we decided to take a couple of days and try to pull off a stunt. Two years earlier, in the midst of record D.C. snowstorms, Senator James Inhofe of Oklahoma had his staff build an igloo on Capitol Hill, with a sign that said, "Al Gore's New House." It was witless (the reason we get record snows is precisely because a warmer atmosphere holds more water vapor—as long as it's below 32 degrees, the flood will be a blizzard) but effective, with remarkable press coverage. We decided to return the favor, and in the brutal heat melt an ice sculpture with Inhofe's most famous declaration—that climate change was a "hoax"—on Capitol Hill.

Capitalism being the strange thing that it is, our D.C. crew quickly found an ice sculptor who could carve the words out of blocks of ice, and deliver them just where they were needed. And so we sent out one e-mail blast asking people to come up with his $5,000 price tag, and raise an equal amount for heat wave relief. Within an hour we had the money, and three times more to help people across the region. We wrote press releases and contacted the U.S. Capitol Police. But at eleven on Friday night I was getting ready for bed when an e-mail arrived from a fellow named Bob Kincaid, a West Virginia activist who'd worked for years to end mountaintop removal coal mining.

Dear Bill,

I hope this email reaches you.

As someone who, like other Appalachian people beset by Mountaintop Removal, has stood with 350's efforts and understands the dire necessity of them, I implore you to PLEASE stop this ice sculpture shenanigan in DC tomorrow.

People in WV are genuinely suffering from the after-effects of a climate-change-driven storm that knocked out power to most of the state last Friday. Tens of thousands of us yet remain without it in life-threatening heat.

That 350 would mock our plight by fund-raising for, and then deliberately melting, a giant ice sculpture on Capitol Hill is the most insincere, elitist blindness to the very real trouble we suffer. Countless Appalachian people would give dearly for that ice; melting it is a direct slap in their faces.

I know the plan is to donate half the funds raised to the Red Cross, but that doesn't mitigate the profundity of the insult.

Our work in Appalachia is hard enough as it is, since we must ever contend with the well-funded coal industry PR machine. Your action tomorrow stands the likelihood of making our efforts here nigh impossible.

It is bad to be down, but we hillbillies are fighting against the ten-count. Please, I implore you, don't let your folks go to DC and kick us back to the canvas. Please.

With best regards,

Bob Kincaid
Board President
Coal River Mountain Watch

That's not the message you want to get before you go to bed. I confess I didn't completely understand the objection, so I wrote him back to say I thought it was too late to stop the plan, and that anyway it wasn't designed to insult—just the opposite. Not everyone supports every project, of course, and so I went off to bed. But something kept me from sleeping very soundly; by four a.m. on Saturday morning I was back at the keyboard and found another message from Kincaid.

> Thanks for the courtesy of the reply. I genuinely appreciate it. I would have not written had the disdain and insult not been real. . . .
>
> These are the people who swelter in misery without electricity, without water and some without food, and have done so for over a week.
>
> How can you hope to have them join this humanity-saving effort after you have mocked them in tomorrow's publicity stunt?
>
> Best regards,
>
> Bob

I sat there staring at the screen. On the one hand, calling it off would be hard. I'd had three of my colleagues—Janina Klimas, Jason Kowalski, Phil Aroneanu—dashing around Washington in the heat for three days, scouting sites and rounding up permits. And we'd sent out an e-mail to every American who supported 350.org—you don't want people thinking you're shaky. On the other hand, I could tell Kincaid's pain was real. West Virginia had been harder hit by the heat, storm, and electric outage than any other state—the state had literally been sweltering. If the criticism wasn't entirely rational, it felt emotionally true.

And I thought about my own emotions for a moment. Calling this thing off wouldn't really damage the cause—indeed, it was important to keep everyone on board, and to respect especially the people living in the hardest-hit places. My reluctance came, I feared, from embarrassment. To *me*. I'd have to say I'd made a mistake—which isn't a very good reason not to do something.

One great advantage of working with younger colleagues is that they're likely to be awake—I called Jon Warnow, one of my original collaborators, at his house in San Francisco, where he was indeed still up at two a.m.; in fact, he was hanging out with Josh Kahn Russell, a newer colleague but one deeply tied into the world of anti-coal organizing. I ran my reasoning by them, they agreed, and so I tapped out a new e-mail blast.

Dear Friends,

I think I screwed up.

Yesterday 350.org sent out an e-mail, telling people that we were going to melt a big hunk of ice in the form of the word "Hoax?" in front of Capitol Hill. We asked for money for it, and also for relief efforts for victims of the heat wave. The idea was simple enough: if this epic heat wave gripping the nation has one small silver lining, it's that it's reminding people that global warming is very very real. And the response was strong—we raised the $5,000 it would have taken to pull off the event, and far more than that for relief efforts.

But we also heard from old friends, especially in nearby West Virginia, who asked us not to do it. The sight of ice melting while they sweltered would be too hard to take; their region, they pointed out, is as hard hit as any in the country by the heat wave, and it would make people feel like their plight wasn't being taken seriously. Bob Kincaid, at Coal River Mountain Watch, said: "Our work in Appalachia is hard enough as it is, since we must ever contend with the well-funded coal industry PR machine." They'd use, he said, the sight of that melting ice to make people feel disrespected.

That makes sense to me.

It's sometimes hard to balance what we should do in one place with what we should be doing around the globe. Climate denial in the U.S. has huge implications for, say, the two million people in Assam, India, currently flooded out of house and home—it's really important to fight people who deny science and hold up needed action. But it's not worth causing trouble to our friends in the process. And the people who fight mountaintop removal in Appalachia are some of our oldest friends; we've been, as it were, up and down the mountain with them. Movements only really work when they move together.

So: no ice melting in DC this morning. We're sending out whatever the reverse of a press release is called. The money we collected will all go for heat and drought relief, and we hope it will do some good. If you'd like your contribution back, let us know (and we'll send a separate mailing to everyone who contributed to make sure they get that chance).

It's been a long, hot, tough week everywhere east of the Rockies; let's hope the heat breaks soon.

Thanks,
Bill McKibben for 350.org

> P.S. The note announcing this thing yesterday came from Jamie Henn, our communications director. But the idea was mine, not his. I'm a volunteer in this effort, and there are days when it definitely feels like you get what you pay for.

By now it was nearing dawn on the East Coast, so once we'd checked with Jason in D.C. to make sure he could actually call it off, Jon pressed the button and sent out the new e-mail blast. It's not fun to tell a couple of hundred thousand people you'd screwed up. On the other hand, one point we'd been trying to make since we'd started 350.org is that we were different—not an environmental organization but a campaign, not a group of slick professionals but a homemade effort that relied on everyone doing what they could. This would be a pretty good test of how far that message had penetrated. (And homemade was completely apropos in this case, since I was still sitting in my boxer shorts at the kitchen table as the sun rose above the line of pines around the meadow.)

The first good omen came from Kincaid. I'd written to him to say that our crew had managed to pull the kill switch at the last minute, and he (probably up at his kitchen table, too) replied immediately:

> Your team has proven nimble, indeed, not only within their prodigious skillsets, but in the quality of their hearts. Nimble hearts coupled to nimble minds are what it will take to prevail in this struggle.

As Saturday wore on, our executive director May Boeve wrote to board members and donors to explain why we'd pulled the plug; one of our communications experts, Dan Kessler, handled media; and my colleague Jean Altomare compiled the responses pouring into our main e-mail dump—1,500 people wrote back, a pretty remarkable tide of feedback. Some, not surprisingly, were upset we'd pulled back; it was in many ways a golden opportunity missed.

> "The sight of ice melting would be hard to take"? Seriously Bill, you couldn't get past that simple means to the end? The message and the media picking that melting ice up would certainly draw more attention to the issue—especially to the deniers. How in the world could you let this one slip by you, and not explain to the Virginians that there are ice caps melting too and this was meant to help stop it?

> I don't get why viewing a symbolic hunk of melting ice would make them feel dissed. It is for them and for all of us that we do creative CD. Oh well. Onward.

> The idea was great—the screw up was pulling out. I'm really sorry to say it.

> Please don't get weak on us and fold to silly things like this—honestly, "The sight of melting ice . . ." It really seems that something else is going on here because how could you not get people past that one. That kind of caving vs. the fossil fuel machine? I'm actually concerned over this cave-in, seems we stand less of a chance against them now.

But Jean said about three-quarters of the responses were positive. We'd offered to return the money they'd donated; almost no one took us up on the offer, and many sent new contributions to be used toward heat relief in West Virginia. People were not necessarily convinced we'd made the right call, but they seemed to appreciate the gesture toward movement—toward the idea we needed to go forward together.

> Compassion and integrity are priceless in this broken-hearted world, and are a part of the treasure of humanity that will see us all through this hard time.
>
> One of my favorite activist emails ever! You guys rock!
>
> Couldn't put up any of the money, but I respect y'all for reversing the intention and considering the effect of this. The reversal could have a bigger impact than the ice melting!
>
> It meaningfully increases my faith in an organization when its leadership truly listens to those around them and can admit with clarity and honesty what they now see as a mistake, even when it means changing a publicly announced course of action.
>
> I'm sweltering in Berlin along with all of Europe and I admit this looked pretty wasteful and insensitive all the way over here, tho I understood where you were coming from. Hang in there and I'm glad you could switch gears at the last minute. Ooops, another blast of heat lighting, signing off quick.

I probably should have been able to figure out in advance that the ice stunt was a bad idea. But I was almost glad I hadn't—by day's end the movement felt stronger, not weaker. If anyone had been laboring under the delusion that we were an infallible team of super-organizers, they now had a clearer sense of the truth—and with it the knowledge that they were going to have to do most of the work going forward. Since I was gearing up in my mind for the big push around the new math of carbon—I had begun to hope it might turn into our most important campaign yet—that heartened me. We'd have no honey today, but we could look forward to strong colonies come spring when we'd need them.

Oh, and I especially enjoyed the reaction at the main climate-denier Web site, run by Anthony Watts and called Watts Up With That? Normally devoted to spreading every pseudoscientific talking point discounting physics and chemistry, its devotees spent much of the day taking me to task. "Moron," "fool," "low-life jerk." My favorite remark, though, came late in the evening, near the bottom of a comment thread:

> They collected money specifically for the ice sculpture. How are they allowed to reappropriate those specific funds for something else? They should be returned to the donors or else it should be investigated as fraud. If they're allowed to use designated donations for "whatever" they think is suitable, they could be using their "Hoax?" money on hookers and blow.

And with that I made my way to bed, to dream of swarming bees and dripping combs. The learning curve is still pretty steep in this movement business; you don't know how the story's going to come out. If the stakes weren't so high it would be kind of fun.

6

. . . . .

THE WISDOM OF THE HIVE

First of all, I got to drive the tractor. In case you think that there's any essential difference between a nine-year-old boy and his gray-haired fifty-one-year-old descendant, forget it. I mean, I got to drive the tractor.

I'd been dropping hints ever since Kirk bought his shiny orange Kubota (one of the constants of rural life is the color scheme of your major implement makers—John Deere, green with yellow wheels; International Harvester, red; Caterpillar, yellow) but there'd always been too much bee work to do. By late July, though, the hives were mostly on their own. Each colony had a super or two on top where the bees could store excess honey—until extraction started in mid-August it was mostly up to the weather, and the worker bees patrolling the fields looking for nectar.

This meant Kirk had some time to think about the rest of the farm—really for the first time, since the summer before had been one long construction project. Now that the house and barn were built, he could contemplate the fields. Which, as fields will do, were growing high with grass. Vermont's soils, even in the Champlain Valley, are not especially fertile (that's why the state's population dropped in half when the Erie Canal provided an easy way to get to the Midwest and its deep black loam); you can raise corn or wheat or, in the best spots, row crops, but the thing you really want to grow is grass. If you don't have animals pastured on the land, you need to cut the hay; indeed, the cycle of the agricultural season moves through first cutting, second cutting, and, in a good year, third cutting. The weather report on the radio every day forecasts the predicted rainfall for the next three days—you need that solid stretch of clear weather to let the hay dry in the field before you bale it, or else it will rot. And bale it is what almost every farmer does—if you don't have your own cows to feed it to, you can sell it to someone who does. "I could let someone come in and mow these fifty acres, and they'd pay me a few thousand dollars for the grass," said Kirk. "But for now I want to return those nutrients to the soil." He was planning, in other words, to do something kind of odd in these parts: simply cut the grass and let it lie where it fell; he'd take the few thousand dollars in the form of a healthier field, which would increase his options in the years ahead.

I live fifteen miles from Kirk—but a thousand feet above him, which in this part of the planet is a world away. I've spent my life in the forest; I've driven by mown fields my whole life, but my clock is set to sugar season, blackfly season, and the orgasmic autumn color. So it was novel fun to stand on the back of the tractor and watch Kirk mow for a few rows—to see the grasshoppers jumping by the thousands, and the occasional field mouse or vole or skunk scurrying out of the way. I got off and walked ahead a few times—the grass was over my head and I'm six foot three. Since he wasn't trying to produce the maximum amount of winter feed, Kirk had waited weeks longer than the other farmers to mow, and the grass was so thick that at points he had to back up and go over the same patch twice. "Now you know why they needed a special plow to cut through that prairie the first time," he said. "This stuff is strong." (In fact, John Deere himself grew up in the town next to mine; he learned the blacksmithing trade at a shop in downtown Middlebury, and took his craft with him to Illinois, where he designed "the plow that broke the plains.") After a couple of passes, Kirk turned the tractor over to me and headed off to his vegetable garden.

In certain ways this was just a bigger version of what you see in a million suburban yards, where the squire on his lawn tractor cuts his weekly swath. The clutch was a little more complicated, you had to raise and lower the mower off the power take-off, and you needed to keep glancing forward and backward to make sure the hay wasn't bunching as it fell—but it was the same basic idea. When you're mowing a lawn, though, the whole idea is negative; you're removing something you don't want. When you're in a field, you're playing with energy—with fertility, with potential. Take it and make some money; leave it and there could be more money in the soil if you can figure out how to harness it. The sun shone hot and the afternoon wore on, and I was totally absorbed, just me and the column of tall grass disappearing as I drove by. I was sorry when the last stand fell—not an emotion I ever remembered in a life of lawn mowing. But I shut the orange beast down and ambled the few hundred yards over to the vegetable patch by the barn, where Kirk was still at work. The cherry tomatoes were ripe and the string beans were ready, so we picked as we talked.

"Now that I've mowed it all once, I feel like I know the farm better," he said. "I'm learning where the grass comes in thick—where it's wet, where the soil is rich. And all the time I'm thinking about the shape of the farm I want to build." As we hunched our way down the rows of peas, he described the vision that was growing in his head for these fifty acres. He planned first to run some beef cattle, once he'd figured out a steady water source and had time to build some fence. "The relationship between cow and grass and soil," he said, "that's what we need to figure out. The energy that grass is trapping from the sun has got to be the basis of things." He sketched out a series of islands amid the fields, which he'd plant with perennials—nut and fruit trees, berry bushes. In between, the cows and eventually sheep and pigs would graze on the grass. As they ate it down, he'd broadcast clover on the grazed patches to diversify the field (and to please the bees). When the apples fell, he could turn the stock into the islands for a while; the manure would always be returning fertility to the soil. "I want to move from a system that works off sun and petroleum to one that works off sun and animals," he said. "I'm glad I've got the tractor, and I imagine I'll have it the rest of my life, but that's the direction I want to go." I could see him imagining a team of oxen coming through the field, hay rake behind them. For me, though, I was deeply happy with the tractor.

• • •

The morning I'd gone out to work with Kirk, *Rolling Stone* published the longest magazine piece I'd written in years. Called "Global Warming's Terrifying New Math," it laid out the argument I described earlier—the stuff about how 565 gigatons of carbon would take us past two degrees, but the fossil fuel industry had 2,795 gigatons on hand. The editors didn't make me pull any punches. "We have met the enemy and they is Shell," I'd written, and they hadn't blanched—in fact, they'd even designed a big chart titled "Enemies List," showing all the major fossil fuel players and how close each of them was bringing us to the edge. I was glad to get it in print; I wanted to lay down a marker, something we could refer to as we campaigned in the year ahead. But I didn't expect much of a reaction.

For whatever reason—the crazy weather, most likely—I was wrong. Not since *The End of Nature* had I struck quite the same kind of nerve. Within a day five hundred thousand people had read it on the Web, and the numbers kept climbing way past two million. By the time the week was out, it had been "liked" 100,000 times on Facebook—since I'm not a Facebook user, that didn't mean much to me, but it clearly elevated my stock with my younger colleagues. Daniel Kessler, an ace part of our communications staff, wrote to say that ExxonMobil, the top company on the Fortune 500, had but 8,800 "likes." Twelve times more likable than ExxonMobil!

As a writer, I found the reaction fascinating—the piece was long and technical, and it broke most of the rules of "messaging" that communications gurus were always laying out. But I'd always found that people can deal with reality—we'd named 350.org after a scientific data point, after all. And we're in an interesting moment journalistically, when the intersection of old media such as *Rolling Stone* and new media such as Facebook can combine to move something across the culture in ways neither could have done by themselves. More to the point, it made me think we had a chance as we looked ahead to the fall—we were making preliminary plans for our post-election road show, laying out the twenty-five cities, trying to line up musicians and speakers. The e-mail was piling up—people wanted to help with the divestment campaign on their campuses and in their churches. The actor Leonardo DiCaprio tweeted the piece out, and suddenly I had a skein of fourteen-year-old girls eager to pitch in. Brave new world!

And, of course, dangerous new world. I got another e-mail in the midst of all this, from my friend Jason Box, the scientist who has spent more time than any American of his generation up on the Greenland ice. He was in Reykjavik, Iceland, fresh back from weeks on the glacier, and he sent a graph of the albedo, or reflectivity, of the ice. We know snow is white—when sunlight hits a glacier, most of it bounces back out to space, instead of being absorbed, as it is by dark blue ocean or green forest. But not all ice shines with the same brightness—even before it melts, warmer snow crystals lose their jagged edges, becoming more like spheres, which reflect far less light.

"You can see it with your naked eye," Jason said when I called him. "Think of the way wet sand is darker than dry." Fresh snow bounces back 84 percent of the light that hits it; even before it melts, rounded crystals can reflect as little as 70 percent. Pure slushy snow saturated by water—which gives it a dark-gray cast, or even a bluish tint—is as little as 60 percent reflective. Add dust or soot impurities and the albedo drops below 40 percent.

Jason's satellite data has shown a steady deterioration in

Greenland's albedo in the past decade, from a July average of 74 percent when the century began to about 65 percent in 2011. And then came 2012—suddenly the line on the chart dropped right off the bottom, showing that at certain altitudes the albedo had fallen by four percentage points in a single season, down to 61 percent.

"I confess my heart skipped a beat when I saw how steep the drop was," Jason told me when I called. "I thought it meant the satellite sensor might have degraded." Instead, weeks of "ground truthing," or collecting data on location, showed it was the ice sheet that was degrading—that the heat accumulating in the ice sheet year after warm, sunny year was making it easier and easier to melt the surface. It's also ominously possible that soot from wildfires in Colorado and Siberia, themselves spurred by climate change, were helping to darken the surface of the Greenland ice—so far he hadn't been able to raise the funds to send a graduate student to do the sampling that would provide the answer.

Looking at the numbers, Jason had written a paper predicting that within a decade the whole sheet might start to melt at once—instead it actually took just a few weeks. Two days after he'd e-mailed me from Reykjavik, NASA satellites showed that at least for a few hours the whole surface of the world's largest island had turned liquid, as temperatures atop the ice sheet reached record levels. "Greenland Melt Baffles Scientists," the *Wall Street Journal* reported. But, in fact, scientists weren't baffled—they knew only too well what was up. "Greenland is a sleeping giant that's waking," Jason said. "In this climate trajectory, the ice sheet is doomed—the only question is how fast it goes."

That fact matters for every corner of the planet, of course.

Water pouring from the Greenland ice sheet into the North Atlantic will not only raise sea levels; it will probably also modify the weather. "If the world allows a substantial fraction of the Greenland ice sheet to disintegrate, all hell breaks loose for eastern North America and Europe," Jim Hansen told me. He'd been talking about this very possibility just a few weeks before, when we were powering through the Sea of Marmara back to Istanbul.

But the future, as pressing as it is, sometimes gives way to sheer awe at the scale of what we've already done. Simply by changing the albedo of the Greenland ice sheet, Jason calculated, the island now absorbs more extra energy each summer than the U.S. consumes each year—the shape and color of the ice sheet's crystals are trapping more heat than all the cars and factories and furnaces in the world's biggest economy.

I didn't know, in other words, whether we'd waited too long for divestment movements and road shows. I didn't know if the fields of grass in the lovely Vermont summer we were enjoying would soon look more like Iowa's parched deserts. I just knew—buoyed by all the people who wrote to say they were up for it—that we were going to fight.

Kirk and I were sitting on an old slab of marble, eating lunch above the waterfall that dominates the center of Middlebury. It's a creamy, churning cataract, maybe eighteen feet high. "For the first settlers who came through here, that must have looked like an atomic reactor would to us," he said. "Endless unlimited power. That's why they made a town here." At the moment, it's used differently—when college is in session, sometimes you'll see kids taking their kayaks down the plunge. An art gallery and a high-end restaurant hug its edge in a postindustrial embrace.

But here's the good news—the old penstock that sent water through a spinning wheel is still intact, and within a year or two local officials hope the waterfall will, for the first time in a hundred years, be generating electricity again—enough to power the entire downtown. After a long era of getting big and distant, our economy, and maybe our culture, has started to make a halting turn toward the small and local. Some of it is jazzy, high-tech. Half a mile west, the college had just finished installing a "solar farm," forty big panels on steel stalks, with motors that reorient them toward the sun throughout the day. A couple of my former students were on the team that negotiated the contract; a local company, AllEarth Renewables, did the work. There's a path that winds through that installation, en route to the college's farm garden, which I helped start more than a decade ago; it's growing steadily bigger, and another of my old students, Sophie Esser, had come back to campus to manage the operation.

I first got to know Sophie almost fifteen years ago, when I taught that college course on local food production, the one where Kirk came to lecture. We spent much of our time planning a garden for the college. It wasn't a plan embraced by every administrator. "We wouldn't want people thinking Middlebury was an *agricultural college*," one dean told me, which I thought was kind of funny, since the college was happy with people studying every other form of culture on the planet. But we persevered—in fact, we broke ground without official approval, but with the chair of the board of trustees picking rocks from the first furrows. In the decade since, the garden had become both beautiful and beloved. Student volunteers under a green-thumbed garden steward named Jay Leshinsky had grown ever more bountiful beds of vegetables, and the garden became a steady supplier of food not only for the dining hall but also for the development office, which discovered that alumni loved the idea of food raised by students—at reunion time the garden was a favorite stop for tours.

When Jay retired last year, Sophie returned to campus. She'd helped start the small farm (planting a children's garden one summer, with kids from local elementary schools), had gone on to earn a master's degree in "gastronomic culture and communication" from an Italian university, and helped run her family's Napa vineyard. Now she was back in Vermont with her seventh-generation Vermont farmer husband, her two-year-old son, and her flowing blond hair tucked up under a Johnny's Selected Seeds ball cap. And she was full of plans: a barn that could do double duty as a classroom (and triple duty as a kitchen for hosting dinners of farm produce), new orchards, a meditation garden, and maybe chickens. Middlebury was getting ready to launch a food studies minor; there was talk of converting one dining hall to strictly local food. All were exciting developments—a recognition from a school renowned for its international studies program that the world was localizing at least as fast as it was globalizing. Two hundred years ago, when Middlebury began, its mission was to give some polish to people who knew how to grow their own food; it was apt that the school was now eager to take our highly polished suburban youth and give them a little grit.

• • •

If you want to think about the way farming has changed, and the way it hasn't, there's no better place than a fair. The drought had dampened the annual festivities across much of the grain belt—in Wisconsin, the pigs at the state fair were about fifteen pounds lighter than usual. "The heat is affecting their virility and appetites," one farmer explained to a reporter. "We've had a hard time getting them to eat enough to get that condition on them." The heat had sapped vegetables and flowers, too—at the Dane County Fair outside Madison, there was only one gladiola in the whole competition. The head judge said that most years when she asked children to describe their projects, "they usually say what they liked best about the plant." This year, "the first thing they mentioned was how much they watered them." At the Johnson County Fair in Iowa, the *New York Times* reported, "attendance fell, four rabbits perished in heat that passed 100 degrees, and a beloved, final fireworks display was canceled for fear of setting off a fire in the bone-dry county."

But in Addison County, Vermont, the annual Field Days, the state's largest agricultural fair, were in full swing this second week of August. As usual I'd circled the dates on my calendar months in advance. Diane Norris, who has directed the fair for the past twenty-eight years, told the local paper, "No matter how many times you see it, every time you come it's new again." I hate to say it, but I think she's mostly wrong—the biggest reason I like the fair is that it repeats with metronomic precision. The fewer changes the better. (For instance, there were new rides this year. My old favorite—a Gravitron remake called the "Starship 2000," a bucket of bolts that dated from some distant past when 2000 seemed impossibly far off—had disappeared. Miffed doesn't begin to describe my reaction!)

So I headed straight for the 4-H building, where I hoped the projects on display would be suitably musty. Sure enough, the first poster board was devoted to "Making Clothes from Scratch," and the second compared culinary outcomes across a wide variety of cookie sheets. ("The aluminum cookie sheet came out the worst.") There were homemade fishing lures, "pillow pets," an illustration of the various parts of a Belgian draft horse, and the classic display of Different Sized Eggs, from the quail on up to the goose. I read about "equine leg protection," how to treat colic, and "All About Pigs" ("pig food are called slops"). Next door, in the Center of Progress building, a man sold Electrolux vacuum cleaners, the National Guard recruited new members, and the local Bible church offered a test to determine "Are you a good person?"

Down in the horse ring the judging was under way in Class 27, Western: "All walk. All canter. Now reverse and trot." The afternoon stretched hot and still, and I could feel the thunderstorm building beyond Snake Mountain a couple of miles to the west, so I made my way to the Maple Building for my annual maple milkshake. (As much as I appreciate those hardworking bees and their honey, I'll always be partial to our crop of syrup from the high-mountain sugar bush.) Next door I bought a fried potato twist ("Don't Be a Hater, Eat a Tater"), and headed into the cool of the old-time barn to look at the antique tools: the corn fork, the hay knife, the slater's hammer, the lard squeezer. And when the rain finally came, fat-dropped and warm, I retreated to the dairy barn. The 4-Hers had shown their prize cows that morning; now the heifers mostly lay peacefully in the hay, and the kids slept peacefully curled up against their warm, breathing flanks. The names were over the stalls: Clementine, Calico, Brownie, Butterscotch. As I said, the pleasure comes from the sense that the real world goes on underneath despite all.

Since I was in fair mode, I headed north a day or two later for the annual meeting of the Eastern Apicultural Society, drawing beekeepers from around the region to the University of Vermont. There was the requisite trade show—bee-shaped travel pillows, hive-shaped teapots, "sting-gel wet wipes," custom bee veils ("We've never had a bee in our bonnet"), electric smokers that "work like a flashlight . . . so you'll never have to light your smoker again." This last item wasn't selling so well—in fact, the six hundred attendees were toting official conference bags made of burlap, perfect for cutting up and burning in their old-fashioned smokers once they got back home. Beekeepers tend to stick with the tried and true—the basic hive box with its ten frames spaced three-eighths of an inch apart was first devised by the Reverend L. L. Langstroth in 1852, and it is essentially unchanged. The smoker dates from the same period, as well as the centrifugal extractor, and the standard technique for rearing queens.

What's new, of course, are the mites and other diseases that have plagued beekeepers in recent years, dangers that led beekeepers to look for help from any quarter, and some of this assistance could be seen on the convention floor. You could buy Apistan miticide strips, or CheckMite, or Terramycin Premix. But if you use the last of these, you're "required to remove the dust six weeks before honey flow," and if you want to buy Hivastan gel, "you must be in an approved state," and if you go with GardStar "yard drench," you "must not use inside the hive!" and you best keep all domestic animals, not to mention "aquatic life," at a distance. Not only is this stuff expensive and noxious, it's increasingly worthless. "The bees keep evolving resistance—it's a real chemical treadmill," said Tom Seeley, the Cornell professor whose book *Honeybee Democracy* I quoted earlier. He's a friend of many beekeepers, who kept coming up to shake hands, and an old admirer of Kirk's. "He really did pioneer this chemical-free beekeeping," said Seeley. "He's like a mutation. Really, in his whole approach to agriculture. He wanted to see if he could build up a business without taking on debt, and he did. And now by going without chemicals. We see untreated bees like his evolving resistance to mites."

Even the majority of beekeepers who do use chemicals are increasingly employing "soft pesticides" that require more skill and effort to apply but leave less residue, Seeley said. The big commercial apiaries, trucking their hives by the thousands to California for the almond harvest, aren't a part of the revolution yet (Seeley: "There's a lot about big-time beekeeping that isn't bee friendly"), but thousands of new beekeepers were joining the fraternity, part of the same locavore wave bringing chickens to suburban America.

I kept running into beekeepers from New York—New York City—at the Vermont meeting. Wally Blohm, who'd been keeping bees in Floral Park, Queens, described the explosion in hives since the city relaxed its laws two years before, permitting rooftop colonies. "I do classes all the time now. People have them on rooftops everywhere," he said. And when they swarm, watch out! After the city legalized beekeeping in 2010, "everybody and their mother" took up the hobby, a city official explained to the *New York Post*—even the Waldorf-Astoria put hives on the roof. But plenty of people got bored, or didn't know how to prevent swarming, hence the mother and the baby trapped in an SUV for two hours because of an Upper West Side swarm, or the swarm that delayed a Delta Airlines flight at La Guardia, or the guy in Queens with three million bees in his backyard, which as NBC noted was more than the human population of the borough. Despite the boom, the bees seem able to find plenty of nectar. "So many people in the city have backyard gardens, flowerboxes," Blohm said. "If there's something there, they'll find it."

Seeley was listening to our conversation. "The hive sends out hundreds of scouts," he said. "It's random. And most don't find anything. But all it takes is a few to come back with good news, and then they can concentrate their labor on that particular place. It's a great example of collective intelligence—it's what we mean when we talk about the 'wisdom of the hive.'"

In this case, the wisdom of the hive seemed to be moving in Kirk's direction. One of the biggest crowds of the whole convention crammed into the university's largest auditorium to hear a talk by Warren Miller, the president of the Pennsylvania State Beekeepers Association, on chemical-free beekeeping. He described the halcyon days before mites ("It wasn't even beekeeping then—it was more like bee having"), and he told of his early experience with pesticides. "Like everyone else I jumped on the chemical bandwagon," he said. "But I was still losing colonies. It was pretty clear I was just breeding stronger mites." So he went cold turkey—he stopped buying new queens and started raising his own, selecting the most resistant colonies for their genetic toughness. "Queen rearing is the pinnacle of beekeeping!" he said to great applause. "Everyone should do it!" He reported proudly that he was now losing fewer than 20 percent of his colonies each winter. Foot-stomping applause!

And where was Kirk? He was at work, forty miles south of Burlington and a mile west of the county fairgrounds, at one of the beeyards where he raises nucleus colonies. These hives weren't for honey production; they'd be sold the next spring to hobbyists from around the Northeast for $220 apiece, and he had the potential to earn something like $30,000 from this small yard. It was a 4-H project made real, and therefore could be just a little stressful. As we opened each hive, he'd check to make sure it was producing brood and storing enough honey to overwinter successfully. "Ooh, this is as good as we'll see," he said—and then the next one was just as lovely, with fat white honeycomb bulging from the edges. "They don't get much nicer than this," he said—but the next one was, if anything, nicer. "Oh, we're on a roll here." Eventually we reached a struggling hive, which he decided to combine with another, which meant getting rid of one of the queens. He found her after a moment's search through the frames, and pinched her dead between thumb and index finger. "Sorry," he said. "This is the playing-God part, I guess."

We got through the last hive. "This yard is set now for months," he said. "I shouldn't need to come back, which is good, because the honey harvest starts next week." I lay back in the sun in my bee veil, listening to the thrum of the bees that filled the air, and behind it the background note of the crickets' song swelling and falling: the comforting sound of life on automatic, the planet working as it should.

The honey harvest started right on schedule at the end of August, but not without at least a hint of trouble. Just before Labor Day, the state health department reported the first death in Vermont's history from eastern equine encephalitis. EEE, as it was being called on TV, is a mosquito-borne disease—one that public health officials had predicted, just a few years earlier, would appear in the state if climate change continued unchecked. Mosquitoes love the warm, wet world we're creating for them—if you were watching our planet through a telescope from some other galaxy and trying to figure out why we were changing the atmosphere, a reasonable hypothesis would be that we'd decided to embark on a planetwide mosquito-ranching business. Along with the deer ticks spreading Lyme disease, they were the clear local beneficiaries of our new climate.

Since EEE kills about a third of the people who contract it, local officials were taking no chances: they were about to begin emergency aerial spraying of an insecticide ominously named Anvil. And since the victim had lived on the edge of the county, the southernmost of Kirk's beeyards would be in the line of fire. The authorities promised local organic growers they'd steer the airplanes away from the borders of their farms—if they didn't, the farmers' organic certifications could be revoked. But Kirk's beeyards were too small and scattered, so he wanted to at least collect the honey supers before the spray dropped.

It was a hot day, and the bees were agitated—though probably not because they knew about the planned aerial assault. "I saw some skunk droppings over there," Kirk said. "They come at night and scratch at the hives. They get the bees to come out and then they eat them, which makes the bees mad." One would think. We were pulling the supers off the top of the colonies and stacking them on the truck; bees were everywhere in the air, and if a drop of honey smeared on some surface half a dozen would descend on the slick and clean it up without a trace. It felt slightly out of control, so we worked fast, tying the supers down and hauling them back to the farm, stacking them in one end of the small trailer permanently parked next to the barn that served as Kirk's extracting room. (Inevitably dubbed the "honey wagon.") Inside, it smelled of honey, and without the loud drone of the bees it was easier to relax.

"I built this seven seasons ago," he said of the trailer. "Since I was renting space, I didn't want to convert a whole room of someone else's building; I always hoped I'd have my own land so I could just move it there." So he meticulously laid out the floor plan for the trailer, making sure he'd have just enough room to maneuver between pieces of gear.

To extract the honey, he pulls one of the ten frames from each super and pivots toward a heated, vibrating electric knife he uses to slice through the waxy caps that the bees have built over the honey cells. The wax drops into a tank, and the frame goes down a small conveyor to the right. Once he's uncapped a couple of dozen frames, he walks a few steps and loads them into the extractor—a big drum that operates more or less like that Gravitron that's no longer at the fair, spinning them so fast that the honey comes out of the cells, collects on the walls, and then slides to a sump underneath the drum. When it collects to a depth of about six inches, a float triggers a pump, which sucks the honey up a two-inch clear plastic pipe—it's a tube of gold that crosses the room to the first of four settling tanks. The wax and other debris slowly rises to the top, and the honey flows into the next tank, and then the next; by the fourth, it's ready to eat. Unlike virtually all commercial honey, it's unfiltered and unheated—as pure honey as you're going to find anywhere. Kirk packs it into fifty-gallon drums and sends it off to his main customer, a Massachusetts retailer called BeeUntoOthers.com, which sells it in glass jars for seventeen dollars a pound, a bargain even at that price. If you go to the farm, you can also buy it straight from Kirk in thirty-pound buckets.

When he built the wagon, though, there wasn't a special market for untreated honey; he just took whatever the commodity market was paying. "That first year was an amazing crop—thirty thousand pounds of honey," said Kirk. "Astonishing. But the price was really low—maybe seventy-five cents a pound. Anyway, as I was cleaning out the last tank, I started feeling sick, like I'd eaten something wrong. And by the next day I was in the operating room getting my appendix out. And then it got infected. It was weeks before I was well again. I hadn't had time to sell my crop—and by the time I got out of the hospital the price had doubled to $1.50 a pound. I made enough money that year to pay for the whole wagon." The price had doubled because the FDA had inspected a freighter from China full of honey and found it was contaminated with antibiotics; it sent the whole cargo back to Beijing, which left the nation's little-plastic-bear honey packers desperate for supply. And the incident started educating consumers that honey wasn't just honey, helping set up a new honey economy that's taken Kirk out of the commodity business.

But he's still a farmer. And farmers, in my experience, usually have something to complain about—the harvest has never quite gone as well as it might have in their imaginations. Some of the frames Kirk was pulling from the supers were bulging with honey (the white wax folding off the edges looks like the almost obscene white fat on an aging steak). But others held barely any honey at all. "This is going to be a mediocre crop, I fear," he said. "There just never was a prolonged honey flow. It started off with such a big rush, but then it dried up. We never got quite the rains we needed to really make the plants blossom." The colonies were in good shape—the hot, dry weather had been far easier on them than the previous year's cold rain. And they'd produced more honey, too: instead of 2011's meager 6,000 pounds, Kirk was guessing this crop would yield 12,000 pounds, maybe 16,000. But it wasn't the bonanza he'd been dreaming of.

I don't think it was the money that irked him—he can cover his expenses just selling queens and nucleus colonies, so the honey is . . . the honey on top. It's more like he's an amateur athlete who's prepared for an event all year: trained for a marathon, say, gotten up in the dark to go jogging every morning. And on the big day it goes . . . just okay. Or maybe it's more than that—he's so identified with the bees that he takes their output personally. "A lot of beekeepers don't like extracting," he said. "They think it's boring, it's the first thing they hire people to do. But to get to see the crop move into a new form—it's just so magnificent to think that it's the result of so much work by so many bees." To have sat and watched for hours as bees arrived at the hives with a few grains of pollen tucked in their saddlebags, and now to see oil drums filling up with honey: it is kind of awesome. And it was pure collaboration—Kirk had fed them syrup to get them through the winter, insulated their boxes, made sure they had queens; in return he'd taken the honey they wouldn't need. I thought of something Tom Seeley had said the week before at the beekeepers' meeting: "Sometimes people say this is the ultimate in capitalism, the bees do all the work and the keeper owns the means of production. But the beekeeper is also a guardian angel. He takes the surplus. But if there's a big rainy spell in September that would starve a wild hive, well, the beekeeper can step in and give them the help they require." The odd bargain between wild and domesticated has its sweet payoff as the harvest starts, and who can blame Kirk if he wants to hit the full-tilt lights-flashing Vegas jackpot? "I just hope global warming hasn't ruined the possibilities of a run of good honey years," he said. "I so enjoy producing those crops."

And having figured out, painstakingly, a system that makes him a comfortable living, he enjoys equally, I think, spreading the gospel. In the rush of the first year's work on the new form, his plan to take on apprentices hadn't yet happened. And he was picky, looking less for enthusiastic new college grads than for people who would understand the commitment required. A few days earlier, a pair of young Amish men had ridden bikes over from New York State for a visit, intrigued by what they'd heard about his operation. "It was good fun to see them roll up, on modern bicycles, with their straw hats and suspenders," said Kirk. "One was sixteen—but since they stop school at eighth grade he's already a full partner with his father in their dairy. And he has a trapline, getting pelts—he works that on his bike." His friend was in his early twenties, and he was the one interested in bees. A few of his Amish neighbors kept hobby hives, but they didn't think an apiary was profitable enough to be a full-time business, especially since without trucks you couldn't cover a wide territory; Kirk's example had lured him over to see. "With the usual method, the most you could produce from one beeyard was maybe seven thousand dollars," Kirk said. "But now, if you're raising queens and nucleus colonies, the theoretical maximum is more like thirty thousand. You won't actually get that much most years, but it changes the math."

Being hard-nosed is what farming requires—the Amish have been America's only consistently profitable farmers. "I've got pretty much the same approach in a lot of ways," Kirk said. "There's the focus on being productive, and there's also the job of keeping expenditures low. It's half and half." He'd clearly enjoyed the visit. "They're not cloistered or closeted minds," he said. "They were both doing a lot of reading, a lot of thinking about things." The difference, of course, is that they had a supportive community around them, who understood exactly what they were doing: a table full of people at every meal and farmer neighbors to call on. Kirk has mostly been alone. But with the honey flowing golden through the tubes it's hard not to be in a good mood. Kirk looked up at me and grinned. "After I do this for a week or two, honey doesn't taste too good anymore. The first year I did it, I didn't get halfway through the harvest when I said to myself, 'This honey tastes terrible. No one's going to buy this.' I rushed out to take it to someone else to taste. 'What's wrong with this?' I asked. 'That's the best honey I ever had,' she said. And she was right—a few days after the harvest was over I could taste it again, and it was delicious."

I drove north from Kirk's house to Burlington for a meeting with Vermont's stalwart senator Bernie Sanders, who was hosting EPA administrator Lisa Jackson. It's not like I've met many cabinet-level officers, but it's hard for me to imagine liking any of them as much as her; an African American woman from New Jersey, she'd fought tirelessly not only against polluters, but against those within the Obama administration who were only too willing to give the fossil fuel industry what it wanted. Sometimes she'd lost—the fall before, while we were in jail, Obama had shamefully postponed a new smog regulation after utility executives had come to the Oval Office and shown him the congressional districts where they'd clobber him with ads if he approved the law. And just in the past week, while Kirk was starting the honey harvest, the U.S. Court of Appeals for the D.C. Circuit had struck down EPA rules aimed at limiting pollution by coal-fired power plants—in the minutes after the ruling, coal stocks had soared.

Jackson sat graciously and listened to many of Vermont's leading environmental officials bring her up to date on state plans for reducing carbon emissions. Vermont's doing nowhere near enough, but quite a bit more than most places. When it was her turn, she spoke bluntly: unless something extraordinary happened in the upcoming election, Congress was unlikely to approve sweeping legislation on climate. The judiciary, filled with Republican appointees, was more and more hostile. Action would come at the state and local level; the EPA would do its best to hold up good local examples and spread them around the country. She sounded tired—not defeated, but a little sad. By any measure the country's environment was unraveling: that morning had come word that the drought had grown so deep that barge traffic on the Mississippi had ground to a halt. (*Bloomberg Businessweek* reported that a Burger King restaurant submerged in last year's flood was now poking up above the waterline.) And yet Jackson spent most of her time in front of hostile congressional committees, battling House leaders who refused even to acknowledge climate change.

I felt—as strongly as I had all year—the desire to retreat to precisely the local work she described; hell, I'd been reluctant to leave Kirk's barn, warm with the smell of flowing honey, to come to the meeting. But they call it global warming for a reason—if we can't figure out a way to tackle it at the top, we haven't got a chance. If Washington was closed off—and our experience with Keystone made me trust Jackson's judgment—then it was all the more important to go after the real problem, the fossil fuel companies themselves. In a way, every time I went to D.C. I felt like I was visiting the cashier at the front of the store. That's the obvious place to start when you've got a problem—maybe she can solve it for you. But if not, going to her for help year after year is just perverse; at a certain point you've got to take your problem to the manager in the backroom and demand what you need. Congress is the cashier. Exxon-Mobil, the Koch brothers, and Peabody Energy are the big boys. That's who we were gearing up to go after now.

We weren't, obviously, going to outspend them. We would need to find other currencies to work in—passion, spirit, creativity. We'd probably have to put our bodies on the line. But what we lacked in cash, we could make up in numbers—that's what organizing was about. If enough bees could fill a fifty-gallon drum with honey, it was worth a try.

As we began to gear up for the road show we'd launch the night after the election, everyone had a job. Someone was booking venues and trying to persuade musicians to pitch in for a night; someone else was designing logos; there were pamphlets to print and videos to shoot. My main job was writing the script for the evening—to take the math I'd laid out in that *Rolling Stone* piece and try to turn it into a bit of theater, which I knew, after years of speaking, meant getting a little personal. You need to offer up a bit of yourself; night after night I'd told the story of how I'd slowly turned from writer to activist, maybe the tale of how I'd come down with dengue fever in Bangladesh and suddenly saw, viscerally, how unfair climate change really was. But the account was always in the same key: I was an accidental activist, making it up as I went along, and kind of sorry to be having to bother anyone; it had the advantage of being at least half true, and it let other people see that they, too, could be leaders. ("If that guy can do it, I sure can.")

So that's how I cast the first draft of the script. But I'm used to speaking extempore; it's rare for me even to have an outline. So normally I can't ask people what they think of a speech until it's done, at which point they may be too polite to say. This time I e-mailed the draft to a few friends to get reactions. The most dramatic response came from one of my oldest friends, someone I'd known since my freshman year at college, and she—quietly—read me the riot act. It was time, she said, to stop pretending I wasn't a leader and to accept that, through no fault of my own, that's what I'd become. The past eighteen months had turned me from a gee-whiz tyro into something else, even if I wasn't quite ready to own up to it. If I was going to ask people to do something hard (and spending a year in a difficult divestment campaign, giving up their time and money and perhaps their freedom for a few jailhouse nights, definitely counts as hard), then I'd better be ready to take charge. "You've always won by being passive," she said. "Right from the start. I mean, the *New Yorker* called you up for a job out of the blue when you were a senior; you didn't have to fight for it. It's time to start taking responsibility."

As I read her words, part of me knew she was right. I resisted for several reasons, none especially noble. For one, as I've said, I'm a bit of a coward. I can usually beat down my fear in the end—in most of the truly important moments of my life, I've actually done the right thing. (When Si Newhouse bought the *New Yorker* and fired the editor, William Shawn, who was only the greatest editor who'd ever lived, I quit the best job in journalism and walked away. So, okay, gut check and I passed.) But day to day I'm conflict averse; out in the tranquil forests of Vermont I can go very long stretches without rancor or bile, and I like it that way. The Internet, of course, means acid can come your way wherever you are, and that's taken a little getting used to. I delete most of the death threats and insane assaults, saving only the ones with some particularly baroque twist:

> You cocksucking, motherfucking, assfucking, Harvard Nazi scumbag moron climatebicile!
> You are the ugliest assfucking moron in the whole rogues gallery. You certainly belong in the new Nuremberg, Pennsylvania Nazi Prison camp for your crimes against humanity!
> Asshole! Shitstain! Harvard Grad!

I knew that the more I identified myself with the battle, the more such craziness would come my way. And it could easily grow scarier, because it would come straight from the fossil fuel industry itself. The year before, a bunch of documents released on the Web by WikiLeaks had shown the oil industry and its front groups dealing with security firms that specialized in trying to "discredit" journalists with every manner of attack, right up to planting fake documents. Not long after that, I had dinner with a friend, an investigative reporter who'd spent years covering the industry. "There is more money here than anywhere on earth," he said. "If you threaten it, they'll come after you in every possible way. If a pretty girl shows up at your hotel door, thank her politely and close it in her face."

But something else scared me more, I think. As long as I was the somewhat bumbling accidental author-activist who'd stumbled into this work, then failure was okay; if you're not expecting to win, then any victory is a bonus. If you're willing to declare yourself a leader, however, then the failure is on you. I didn't care what others thought, so much, but I'd lose my own emotional cover—at some level it would become my responsibility. It wasn't just what I was doing for a few years until normal life could resume—it would be who I was. And since our odds were slim at best (did I mention that this is the richest industry on earth?), that didn't seem so appealing. Still, I'd told hundreds of crowds: "I don't know if we can win. But I know we can change the odds some. And given the stakes, that's worth throwing yourself into this fight." Maybe I should take my own advice.

My clearheaded friend wrote again. No one will blame you if you lose, she said. "No one will see it as anything but the most valiant effort." She quoted from the end of a recent profile of me in *Outside* magazine:

> If, as is far more likely, he has zero impact, and we become Venus 2, and all those pixels of snowflakes and sand castles and little girls holding signs are nothing but melting chips of silicon on a dead server, then it won't be because William Ernest McKibben didn't give it a shot.

She finished her upbraiding like this:

> If you make any mental adjustment, I wish it would be to move away from the self-description that you're a "mild mannered Methodist Sunday school teacher." You're not. You're a fighter and a hero. That's how others see you. So really the only one you're now fooling with the Sunday school line is yourself.
>
> Maybe it's time to shift from being Clark Kent to being Superman.

Superman was clearly beyond me—I mean, I'm afraid of bee stings. But it was probably time to stop being Jimmy Olsen anyway.

All this dithering got less theoretical over two days in Brooklyn, in an upstairs room overlooking a gelato factory and a Hasidic-owned auto body repair shop not far from Myrtle Avenue. Such was the headquarters of International WOW, the theater company formed a decade earlier by a young refugee from rural Pennsylvania named Josh Fox who'd come to the city, gotten himself a pair of those cool round glasses and a Yankees cap, and become an early (and actually cool) subtype of the hipster. Actually, he was not a hipster, because he cared a lot about justice and refused to escape into irony. Some kind of karma gave him his shot in 2008, when a gas company tried to lease the farm he'd grown up on along the New York–Pennsylvania line; he began to investigate this novel "fracking" technology and soon turned into its principal scourge, producing a documentary called *Gasland* that showed the land rape and water pillage that followed the industry wherever it went. Our paths inevitably converged, as I started to understand the climate implications of a huge new store of hydrocarbons beneath our soil and he began to figure out that the fracking wells of Appalachia were just one sign of the drive for extreme energy that was breaking the planet. Mountaintop removal, tar sands mining—and ultimately the global warming they all fed—were the enemy; fighting them one by one wouldn't work, and so he'd come to D.C. and gotten arrested in our Keystone fight, just as 350.org had helped organize the drive to convince Governor Andrew Cuomo to ban fracking in the Empire State.

Anyway, we were allies. And so he offered to turn his talent for theater to our service, opening up his headquarters for early rehearsals. Fueled by bagels from down the street and soda from the machine at the car rental place around the corner, we got to work. I read what I'd written, standing at a podium, and a mix of environmental campaigners and theater pros offered reviews, hour after hour. Nellie McKay, a beautiful singer with an even more beautiful voice, dropped by for an hour; Omar Metwally, who'd starred in *Rendition* and was about to play a vampire in the next *Twilight* film, stayed half a day.

We worked on lights and cues; we divided the two hours of script into sections; we tried to figure out what images to project and where it made the most sense to stop for music. At every break I'd disappear through the trapdoor in the stage to Josh's office below and tap out a new set of revisions—it's the closest I'll ever get to the experience of a lyricist trying to repair songs out on the road before a Broadway opening. I was coming up with scenes I liked (pouring one can of beer after another into a giant vat to depict just how far the industry was willing to overshoot the carbon-holding capacity of the atmosphere), but I was also fighting my own wimpy demons. The prologue went from me saying, "For the past few years I've been an activist—a reluctant and not especially skillful one" to something a little stronger:

> I feel as if, for me, this may be the start of the last campaign I get to fight. Not because I'm too tired to go on. But because the planet's getting tired; the moment's come to make the stand. We're reaching the limits, running out of time. But that doesn't depress me—I'm more excited than I've ever been, because I think we know what we need to do. I think we've peeled away the layers of the onion and gotten to the heart of things.

That sounded more confident than I actually felt—but maybe that's one definition of a leader.

We agreed to reconvene in mid-October for a grand dress rehearsal, open to the public, in the hopefully friendly surroundings of Burlington, Vermont. And I went out on the road for several weeks of speeches that drew big crowds. I could tell that the *Rolling Stone* article had hit a real nerve, and I was able to test-market some of the ideas for the divestment campaign at Amherst, Madison, Ann Arbor, and New Haven. But I knew that even a huge, enthusiastic crowd guaranteed little; for now, I wasn't asking anyone to actually do anything. That would come soon enough.

When I left home in late September, I knew I wouldn't be back under my own roof for more than a night till mid-December. So in the midst of hectic preparations, I gave myself a vacation and spent a morning with Kirk, helping feed beehives so they'd be set for winter. This was the payback for the honey he'd taken in the weeks before. He'd just finished totting up the season's total: twenty-six fifty-gallon drums at 620 pounds per drum equals 16,120 pounds, or a little more than he'd predicted early in the harvest—at three dollars a pound, which is what he was expecting for his untreated honey, that's a reasonable haul. He'd stirred up a few five-gallon pails of sugar syrup in the barn that morning, and now he was using a galvanized watering can to pour some into the feeding troughs of many of the hives. Not all—some had made so much honey that he deemed them safe from starvation. But the rest were going to need a little help to make it through the seven months till plants began to bloom again. "See those purple asters over there? They'll still bloom even after the first frost. But that's really all they've got left to work with," Kirk said.

It was chilly when we began, though the sun soon warmed past the magic 50 degree mark and the bees began to fly. But the maples were turning fast, with flashes of red all around us. And overhead, above the buzz, I heard a flock of geese trumpeting by on their way south. As he started hunkering down for the winter, Kirk was experimenting with high technology—he had four solar-powered "Nite Guard" lights attached to the four corners of one box. When night came and the sun went down, their white lights would flash irregularly all night, which was supposed to be enough to ward off bears.

"Do you know those Mullah Nasruddin stories they tell in the Middle East, about the smart guy who might be a fool?" said Kirk. "One day a man saw Mullah Nasruddin sprinkling white powder in his garden, and asked him what it was for.

"'Oh, that's to keep away the tigers,' he said.

"'But Mullah, there are no tigers for hundreds of miles,' the man said.

"'Oh, it's highly effective,' said Nasruddin."

As we talked, we poured sugar syrup patiently into the hives. The bees were remarkably calm. "The last generation of bees for the winter is different, though you can't tell by looking," said Kirk. "They don't want to waste any energy stinging unless they absolutely have to. They're very mellow ladies."

The feeling of fall approaching—with hay in the barn or honey in the drum or wood in the shed—is one of the finest feelings there is. It's one of the reasons I've never quite understood the rush for gain and increase. In a linear calendar, the third quarter of 2012 is very different from the third quarter of 2011, and the only acceptable description for that difference is "more." But in the actual, cyclical, seasonal world, the fall of 2012 is much like the fall of 2011, and what you need is pretty much what you had the year before. Kirk handed me a pound jar of pure gold, the first he'd bottled from this year's crop. "This is good honey this year," he said. "There's a lot of basswood in here; it's got a herby aftertaste. So good."

In the new world we'd inadvertently built, of course, the

difference between the fall of 2012 and the fall of 2011 was that in the interim a lot more of the Arctic had melted, and the ocean had grown more acidic, and we'd come through a gruesome drought. So I went back to work, back on the road, but at least I had that sweet taste in my mouth.

7

. . . . .

ADRENALINE AND MEDITATION

The month before the tour started was, in truth, a tour of its own. I crisscrossed the country—from Marin County to the University of Texas, from Syracuse to Columbia, Missouri, from Austin to Boston—and then spent a week in Canada, where a huge movement had grown up to oppose the tar sands pipelines aimed at the Pacific Coast. Five thousand British Columbians gathered on the lawn of their provincial parliament to risk arrest; a few nights later, as I waited to talk to a giant crowd in the coastal town of Nanaimo, Chief Doug White of the Snuneymuxw First Nation welcomed us with a speech in his native tongue. "If any of you didn't follow that," he said at the end, "it translates roughly as 'There's no goddamned way they're building that pipeline out here.'" I made it back to Ottawa the next night, where hundreds of youth climate activists were holding a conference—and where I got to meet Naomi Klein and Avi Lewis's baby, Toma, for the first time. He was gorgeous, alert, and full of smiles.

At every stop along the way, though, I was checking the Internet. About a week earlier, a storm had begun to form in the far Atlantic. At first most of the computer models had predicted it would be sucked to the east, harmlessly out to sea. But there was a small chance it would take an unlikely curve west instead, following a track close to Irene's the year before. Two powerful North Atlantic hurricanes in two years seemed statistically unlikely, especially since this was late October, past the season for the strongest storms—but then, normal no longer works as a way to measure things. When we left the Holocene, we left the predictable: this was already looking to be the hottest year in American history, we'd seen the deepest drought in living memory, and the full-on melt of the Arctic had shocked even the most pessimistic of climate scientists.

And this new storm—now named Sandy—seemed intent on visiting the Atlantic Seaboard, where the water temperature was again five degrees above normal. In Vermont we'd only begun to rebuild. I got to Boston the day before the storm was due to hit, but it was so vast—the largest storm ever measured, with tropical force winds stretching 1,040 miles out from the center—that the outer bands of rain and wind were already whipping the Hub. I talked to a church full of religious environmental activists and then went to City Hall Plaza, where a few dozen stalwarts had been camped for a week, demanding that the Massachusetts senatorial candidates, Elizabeth Warren and Scott Brown, at least mention climate change in their upcoming debate on the following Tuesday, a week before the election. The rain was coming down sideways as I talked through a bullhorn, and then two young volunteers drove me home to Vermont, to help make sure the hatches were battened down; I'd scrubbed a few talks and changed my schedule so I could get home just in case.

Watching from a distance as Vermont drowned the fall before had been hard—the rivers I knew and loved had suddenly turned swollen and snarling, eating away their banks and swallowing houses and farms. It was worse than anything any Vermonter could remember—but it wasn't, somehow, *weird*. It was at the upper end of the horror scale, but the scale was marked in comprehensible increments.

Watching Sandy flood New York, though, was different. It felt scarier by far, like a glimpse into the way the world ends. More people died on the Jersey Shore, on Staten Island, and on Long Island (and in Haiti, where Sandy had overwhelmed refugee camps and sparked a full-on outbreak of cholera), but New York is our national city, and really our global one, too. If you haven't been to Manhattan, you've ridden its subways in a thousand movies and TV shows. Now those subway tunnels were filling with salt water, millions of gallons of cold Atlantic surging down the steps of the IRT and the BMT and the IND. The Lower East Side—stuck in the national imagination since the days of Jacob Riis—was suddenly a branch of the East River, whitecaps breaking across its intersections. Sitting in front of my computer, staring at the images accumulating on Instagram, I wrote a short piece for the *Guardian*, just because I couldn't think of anything else to do:

> New York is the city I love best, and I'm trying to imagine it from a distance tonight. The lurid, flash-lit Instagram images of floating cars in Alphabet City or water pouring out of the

East River into Dumbo, the reports of bridges to Howard Beach submerging and facades falling off apartment houses—it all stings. It's as horrible in its very different way as watching 9/11.

But it's the subways I keep coming back to, trying to see in my mind's eye what must be a dark, scary struggle to keep them from filling with water. The tide at the Battery has surged feet beyond the old record; water must be pouring into every entrance and vent—I hope some brave reporter is chronicling this fight, and will someday name its heroes.

For me, the subways are New York, or at least they're the most crucial element of that magnificent ecosystem. When I was a young Talk of the Town reporter at the *New Yorker*, I spent five years exploring the city, always by subway. This was in the 1980s, at the city's nadir—the graffiti-covered trains would pause for half an hour in mid-flight; the tinny speakers would reduce the explanation of the trouble to gibberish.

It was how I traveled, though—I didn't even know how to hail a cab. For a dollar, you could go anywhere. And my boast was that I'd gotten out at every station in the system for some story or another. It may not have been quite true: the Bronx is a big and forgotten place, and Queens stretches out forever—but it was my aspiration.

The subways were kind of dangerous, but also deeply democratic. Writing about homelessness, I slept with hundreds of other men on the endless A train to the Rockaways. I convinced motormen to let me ride as they turned trains around through the City Hall station abandoned decades earlier. I hung out in the control room under Grand Central with its Hollywood array of levers and lights.

> Imagining all that filled with cold salt water is too much.
>
> I'm an environmentalist: New York is as beautiful and diverse and glorious as an old-growth forest. It's as grand, in its unplanned tumble, as anything ever devised by man or nature. And now, I fear, its roots are being severed.

And then, for the next four or five days, I just kept writing. Vermont was basically unharmed—our power flickered and went off briefly, but the backup batteries from the solar panel kept the Internet working, which was mostly what I cared about, because this storm clearly had the chance to change the way we understood the way the planet worked. I couldn't get to New York to help—New Hampshire and New York closed down the highways. But I could raise money for relief, as 350.org quickly partnered with Occupy Wall Street and others for an improvised effort that brought aid to regions the Red Cross and FEMA seemed unable to reach. And I could help explain what was going on: that this was an unprecedented tempest, with the lowest barometric pressure ever recorded north of Cape Hatteras.

But unprecedented, I emphasized, didn't mean unexpected: *this was what happened* when you changed the planet's ground rules, and scientists had been warning for years to expect a cataclysm of this kind, right down to predicting how deeply it would flood the subway tunnels. I wrote for the *Daily News* (recommending that hurricanes henceforth be named for oil companies, not girls) and the *New York Review of Books* and pretty much everyone in between. But even if I hadn't written a word it wouldn't have mattered—the pictures did the work, unlocking in people's unconscious a cascade of images derived from Revelations or Hollywood or some nightmare mash-up. We'd seen worse storms in every corner of the planet—hell, in Pakistan in 2010 the rain and flooding were so intense they chased twenty million people from their homes. But if God's aim was to wake folks up, His aim was improving. As meteorologist Jeff Masters put it in his wrap-up of the storm, New York, the media and financial capital of our home planet, had not "experienced a storm this strong since its founding in 1624." By Thursday, *Bloomberg Businessweek* had splashed these words in huge black letters across its cover: "It's Global Warming, Stupid." That afternoon a shaken Mayor Michael Bloomberg, a political independent, endorsed President Obama in the election then just five days away, citing climate change as the reason. The climate silence, meticulously maintained by the candidates all year long, had been broken by Mother Nature.

Air travel resumed along the East Coast by week's end, and I flew to North Carolina to give a speech to six thousand scientists gathered at the annual convention of the Geological Society of America. They gave me their annual President's Medal, less I think for my work than for the fact that they needed somehow to publicly recognize what their members were able to sense more easily than most: this earth had left behind one epoch, the Holocene, and was now careening into something else. They were watching in real time one of those transition boundaries that only carbon dating and sediment samples had revealed from the deepest past. Viewed that way, our fight to slow it down seemed almost nearly pointless—but not quite. The earth was shifting, but perhaps we could still determine how fast and how far.

And so, on to Seattle, and on to the fight against the fossil fuel industry. I got there a night early, so I could watch the election returns with Sam and Lisa Verhovek, two of my oldest and dearest friends. Sam had spent his newspaper career with the *New York Times* and the *Los Angeles Times* before quitting to write books, so he'd covered many an election; he knew how to read the numbers starting to pour out of Ohio and Florida, and long before the networks called the race he was saying Obama had it won. I cheered—I'd organized the largest protest of his first term, but I knew exactly what Karl Rove and other Republican strategists had in mind for the planet. But ten minutes after NBC flashed "President Obama Re-Elected" on the screen, we pushed the button to send out an e-mail to four hundred thousand people announcing the first protest of his new term, ten days hence, in D.C. We needed to show the president that we hadn't forgotten about the Keystone pipeline. The conventional wisdom said it was certain now to be approved. But not without a fight.

Before we got to that, though, we had another, bigger fight to pick, this one with the fossil fuel industry itself. Our tour would start the next day, and so I headed off to bed.

The minute I walked into Seattle's vast Benaroya Hall, the venue for the first night's show, I was unsettled. I knew we'd sold all 2,500 tickets, but I hadn't really quite figured out how many seats that would turn out to be. I've spoken to larger crowds but always at some convention or festival or rally—these folks in Seattle were coming to see *me*, and suddenly it felt as if I might not have quite enough to say. Jon and Duncan, who were running the visuals, were about a football field away, in a booth two balconies up. We ran through our rehearsal as best we could, and then I went backstage to pace.

The mayor, Mike McGinn, welcomed the crowd—and then he announced that he'd spent the afternoon huddled with the city treasurer, starting to figure out if it was possible to divest the city's own money from fossil fuel stocks. The crowd roared, and then they roared at the four-minute video that started the show. Filled with images from five years of 350.org, it was narrated by Van Jones, who had spent the night before as an election-night commentator on CNN. He'd been Obama's first green jobs czar, but the conservative talk show host Glenn Beck had chased him out of D.C.; now Beck had disappeared in a puff of noxious smoke, and Van was back in the limelight. It seemed a good omen.

I walked out on stage and started to talk—and within seconds a heckler jumped up and started shouting that global warming was a hoax. So much for omens. I was a little stunned; a couple of people surrounded him and started trying to talk him down, and I did my best to make a joke about "even in Seattle," but I was rattled. He was apparently a follower of Lyndon LaRouche, the leader of a marginal but tenacious political cult that for some reason had fixed on me, a "Middlebury College Nazi" and "savage Malthusian" who "demands the takedown of man-created infrastructure on Earth." (I'm in good company, since LaRouche is also convinced that the queen of England is a drug dealer.) The heckler finally strode out the endless aisle, still bellowing, and the evening settled down. We worked our way through the math, and the videos from Naomi Klein and Desmond Tutu, and a few apt songs from a young Native American musician named Nahko. The crowd roared at all the right spots, but I can't recall many moments—just hugs from friends at the end, and then we were aboard the bus.

And the bus was pretty great. Your average fifty-one-year-old book author with a receding hairline doesn't get that many opportunities to feel cool. But this was definitely one of those moments. Our driver used to drive for Johnny Cash; by power of association that's as good as it gets.

We sped south on I-5 out of Seattle, under Mount Rainier, headed for the Oregon border. There were eight of us aboard, which was good since that's how many bunks there were—and better yet there were four WiFi hotspots, so all the hot and cold running Internet you could ever want. Which was also good since we needed it to find biodiesel stations en route. I got the late great Dobie Gray ("Out on the Floor") cranked on Spotify and settled in to write blog posts.

The problem with fighting climate change is that it never feels like we're getting anywhere. Right at that moment, though, we were getting to the outskirts of Portland. And maybe the outskirts of doing some damage to the ExxonMobil mystique, the Chevron reputation, the Shell brand. I had my 350.org baseball cap on, my earphones pulled down tight, and now my northern soul playlist has turned over to the too-soon-forgotten Prince Philip Mitchell and his not-quite-a-hit "I'm So Happy." Don't know if we're going to win, but we were rolling.

I had to leave the bus temporarily in Portland and fly ahead to San Francisco because it was going to be a long day. Wheels down by nine, and into town to be interviewed by former Michigan governor Jennifer Granholm and former San Francisco mayor Gavin Newsom for their news shows, then on to the city's famed Commonwealth Club to meet another former—the ex-CEO of Shell, a fellow named John Hofmeister, whom I was scheduled to debate. Some combination of weariness and cockiness turned me a little more savage than usual—every time he tried to explain that energy was a complicated problem that needed endless compromise and study to solve, I reminded him that he'd been the one applying for the permits for Shell to drill in the Arctic once it became clear that global warming had melted the ice, and that he'd been a big backer of the Chamber of Commerce as it poured campaign money at climate deniers. Brokering some kind of compromise between Big Oil and environmentalists wasn't the point—the negotiation was between human beings and physics, and physics wasn't going to bend. It was a rout, and I left both cockier still and even more exhausted, just able to keep my eyes open for what was actually the day's most important session, a meeting with twenty of the pioneers of social media in a conference room at Twitter. These folks were young and bright, and I'm guessing the average net worth in the room was eight figures—they wanted me to brief them on climate change for an hour, and then to leave so they could break down into small groups and figure out how they were going to solve it via some viral outbreak of interweb platforming wizardness.

The next day started at the Green Festival in San Francisco with a brunch for bloggers and then a talk at one end of the cavernous hall; thousands of people sat on the floor and stairs and hung over the railings. It was a trip to finish with the crowd on its feet and then sprint fifty yards to the bus, which was clearly visible through the plateglass window, and hop on board as it pulled away for the trip to Palo Alto, where I entertained the crowd with the story of my rejection from

Stanford as a high school senior. As we drove all night down I-5, I was lulled to sleep by the vibration of the engine. We got to Los Angeles in time for me to put on decent clothes and head up Mandeville Canyon to what may be the single most beautiful house I've ever seen, the hilltop property of Norman and Lyn Lear, who were hosting a session for fifty or sixty screenwriters in hopes they'd insert story lines about climate change into their films and TV shows. I told them that the moment for screwing in new lightbulbs had passed and that now we needed real activism, and Jim Hansen chimed in via Skype. After questions, I left those heights and descended back down into the city itself, where a packed house was waiting at UCLA for the evening's show. Good fun.

And the next day was our day off—we flew from LAX to the Atlantic, leaving behind our bus, which had developed mechanical troubles. ("It's eating oil," the driver explained.) We'd need to pick up a new one back east. Five nights down, sixteen to go.

I love the West Coast, but I'm at home in the East. It felt right to pull into the Holiday Inn in Portland, Maine, and stumble out, jetlagged, into a cold, foggy evening. We slept soundly, and then got ready for the next night's show, at the truly lovely State Theatre. We'd sold it out weeks before, but I knew it was going to be electric for another reason. Tiny Unity College, in the tiny town of Unity, Maine, is one of the country's greenest colleges. For years its dining halls owed their hot water to the solar panels a professor salvaged from the White House roof, when Ronald Reagan took down Jimmy Carter's early, prescient installation. But its current president, Stephen

Mulkey, had taken things a considerable step further—at my urging, he'd asked his board of trustees if they'd divest. And a day before our show they'd voted, unanimously, to do just that: the first college in the country to join our effort. When I introduced Mulkey and he got up to tell the story, the theater just exploded—it was proof that what we were asking for wasn't impossible. In fact, the more I thought about it, the more likely the whole scheme seemed. There were crowds of students from Bates and Bowdoin and Colby and a dozen other colleges in the hall that night—each of those schools had Web pages boasting at great length about its sustainability efforts and its commitment to greening its campus. All those bike paths and energy efficient buildings were great in themselves—and a logical argument for greening the portfolio. How could you pay for them with Exxon-Mobile money?

There was beer every night—a skit where we illustrated the climate math with me opening a single bottle of some good local brew, and then volunteers illustrating how much the oil companies had in their reserves by carrying up bottle after bottle till our makeshift bar was covered. The metaphor worked pretty well ("Drink this much and you'd be ripped, wasted, polluted, smashed, totaled, wrecked—now think about the planet, and how long that hangover will last"), but it was also a chance every night for the audience to relax a little amid all the tough news. And it was a chance for me to have a swallow of beer. That night, with the Unity news ringing in the air, it tasted even sweeter than usual.

And the next night was pretty good, too—a homecoming of sorts. I'd grown up in the suburbs of Boston and gone to college at Harvard; the city is the capital of American higher education, and so we'd known all along it would be a crucial hub of our divestment campaign. We'd sold out our first venue—an eight-hundred-seat church—in twenty-four hours, so we'd gone looking for a new home. The only place available was the Orpheum, a venerable hulk just off the Boston Common in the center of the city, which began life as a movie palace but became a concert hall in 1971 (with James Brown as the opening act). I'd seen the Ramones there in 1978, as a college freshman. It was dauntingly huge, the biggest hall of the tour—but we sold it out, too, and the place was absolutely buzzing. Thank heaven it was the night that Naomi Klein joined the tour live.

She'd been in on the planning from the start, obviously. But having a four-month-old son does not jibe easily with driving around America in a bus, so we'd settled on back-to-back shows in Boston and New York as easiest for her. The video she'd cut for the other nights was sterling, but having her there in person made life much sweeter—we got to play around with Toma for an hour backstage before the show (life would be better if everyone just went around blowing raspberries on each other's tummies), and then I got to play around with her on stage for two hours. She took half the script, and we threw lines back and forth like pros. It couldn't have gone better, which was nice, since my mother was in the third row. We were pumped up as high as I can recall by the time we finished.

Which sounds great, except that somehow you then have to get to sleep. Adrenaline is your friend—there's no way to engage three thousand people without a steep dose of it coursing through your arteries—and then it's your enemy. For the first time in my life, I have a real sense of why so many musicians end up doing some combination of drink, drugs, and groupies. I did fitfully sleep, and rose at four forty-five for the early Acela to New York—I was going ahead of the bus to talk with Bob Semple, the legendary editorial writer for the *New York Times*, and then David Shipley, who once ran the op-ed page at the *Times* and now does the same thing for *Bloomberg*, and then the ABC enviro correspondent, and then—there's a lot of media in New York.

And a lot of story. We were barely three weeks past Sandy, and though the tunnels had been pumped out and most of the subways were running again, much of the city was still a mess—the Rockaways, Breezy Point, and Red Hook were piled high with rotting trash. We'd been working with our colleagues at Occupy Wall Street since the storm hit—they'd been remade for the moment into Occupy Sandy, so the show that night, in the giant Hammerstein Ballroom, felt different from the rest of the tour. There was more anger, because if anyone had taken climate change seriously a quarter century ago when I started in on it, we might not have set the scene quite so perfectly for this monster storm. The work was more pressing as well—there was a big table in the back with people volunteering for relief efforts.

Josh Fox—the filmmaker whose *Gasland* had sparked the nationwide anti-fracking movement—joined us in person that night. He explained to the crowd how all the local fights need to come together as one. "We've been like the little Dutch boy," he said, "sticking fingers in a hundred holes in the dike" as we fought frack wells and coal ports and pipelines. "We've got to join those fingers together in a fist," and go on offense against the fossil fuel industry. He's right—there's simply not enough time to do it one local fight at a time, playing an endless game of Whac-A-Mole with the endless bad ideas a profit-obsessed fossil fuel industry keeps hatching. Instead, we need to take away their social license, turn them into pariahs, and make it clear that they're to the planet's safety what the tobacco industry is to our individual health. I'm old enough to remember when working for Philip Morris was a perfectly honorable job, back before we knew that cigarettes were killing folks. But once we did know, that changed, and Philip Morris became Altria, in a vain attempt to shed its repugnant image. We've got to do that again.

From New York we traveled to Philadelphia, and then on to D.C., rolling in next to the Warner Theatre in the early morning hours. We had to set up fast, because just this once the show was a matinee, designed to end at three p.m., so that we could spill out into the streets outside for our Keystone demonstration before dark.

As I've explained, we'd pushed the button on that invitation to the demonstration ten minutes after NBC had called the election for Obama. The inside word was that he'd approve the Keystone permit once the election was past, so we thought we needed to strike quickly to show that no one had forgotten. And a day after the election I was glad we'd done so, when the head of the American Petroleum Institute, Jack Gerard, told reporters that the oil industry had been "implicitly promised" by the administration that the project would now be approved; a day later the giant rating agency Moody's said the same thing. Washington wisdom can set faster than superglue, and once it does it's hard to pull apart, so we needed to make it look like a contest from the start. But eleven days to pull together a mass protest is pushing it, even with a crack crew like ours, so I worked extra hard up on stage to make sure the audience was fired up enough that they'd flow straight into the demonstration. And indeed they did, meeting a couple of thousand more who hadn't been able to get tickets. We followed an eight-hundred-foot inflatable pipeline in a long snake dance around the White House; by the time we had reassembled in Freedom Plaza the sun was going down, and the temperature, too, but it was a spirited group—they stood and cheered through the end. Naomi had said it best during our New York show: we were glad that Barack Obama had been reelected, "but this time, no hero worship and no honeymoons."

As the bus rolled on to North Carolina that night, two particularly good omens emerged. The first was on the Keystone front—we'd started that fight with our small ragtag army, but the big green groups had become increasingly larger players. Now the Sierra Club was offering to take the lead for the next big demonstration, one that we were planning for President's Day in February. And that moved the odds a little further; now approving the pipeline would mean not only disappointing the activist wing of the environmental movement, but also its biggest single organization. The Sierra Club under John Muir had saved Yosemite, and under David Brower had stopped the flooding of the Grand Canyon; now, under its new director, Michael Brune, it was taking on this most pitched of battles with the oil industry. Brune joined us in North Carolina, taking the stage at the vast Duke University auditorium to tell the story of the past few weeks of his life. Born and raised in one of those New Jersey barrier island towns, he'd been helping his folks piece their lives back together after Sandy wrecked their home, searching through soaked scrapbooks and mucking out moldy basements. He seemed ready to fight.

The other omen came from the divestment battle. In the ten days we'd been on the road, about one hundred campuses had joined in, with more almost by the minute. Unity's announcement that it would divest had been a jolt—but Unity's endowment was tiny, just $12 million. We received news that more than four thousand Harvard undergraduates, in an officially sanctioned referendum, had voted by a three-to-one margin to demand that the college divest its holdings in fossil fuels. I'd seen a boy on campus that afternoon with a sweatshirt that said "Harvard: Duke of the North."

"Prove it," I told the throng that gathered that night, and they roared back that they were up for the challenge. I knew how long the road from student interest to trustee action would be—but I knew, too, that just getting the battle going was a win for us. If people were talking about the evil oil companies, we were winning.

We couldn't leave for Atlanta right after the show that night—our driver's logbook showed he needed a few more hours of rest before he'd be legal again. I climbed into my bunk and fell slowly asleep, waking when the bus lurched to life a couple of hours later. I groggily checked my watch: sure enough, we were now the midnight bus to Georgia.

It turned out our Atlanta theater was in Little Five Points, a trying-almost-too-hard hipster enclave of skateboard shops and vintage vinyl outlets. I pulled on my suit for a retro moment of my own, a noontime visit with some of the city's key black clergy, summoned by the Reverend Gerald Durley, who'd first met Martin Luther King Jr. in 1959; in fact, the road to our lunch spot went right by King's birthplace and Ebenezer Baptist Church, where he had followed his father into the pulpit. Durley had been an early champion of climate change action; it felt good to be with some churchmen, to say grace over lunch, and to feel the movement broadening a little.

Our theater that night was a Depression-era vaudeville house that had closed in the 1960s to avoid desegregation, and only opened again in the 1980s. It was full to the brim, with students but also with white-haired older ladies, who laughed and nodded when I told them that getting arrested was one of the few things that got easier with age. "Past a certain point, what the hell are they going to do to you?" I said. The crowd rose as one at the end, promising that they'd march as southerners (black ones, anyway) had marched once before. I headed for the airport, home for Thanksgiving for three days, feeling buoyed—we'd done thirteen straight sold-out shows. At the gate I fired up my computer, and there was a special online *Rolling Stone* photo essay about the first half of the tour, sixteen powerful images of the crowds that had gathered. So far so good.

After our Thanksgiving break, the tour resumed in Providence, Rhode Island, and this time it was a little less chaotic—most of these towns had fewer newspapers and TV stations, which was good because the bus rides kept getting progressively longer as we headed west. From Columbus to Chicago, from Chicago to Madison, to Minneapolis, to Omaha, to Boulder, to Salt Lake City—bouncing between red state and blue, between plain and mountain. One day you'd be with ranchers and the next with people who would run screaming from a hamburger—my busmates included, who ate an inordinate amount of tofu; I told them I was planning to open a vegan restaurant chain called Tastes Surprisingly Okay. Eventually, inevitably, the bus became the common denominator—I wrote a short guide for life aboard that went up on the blog:

RULES FOR SLEEPING ON THE BUS

1. Feet go toward the front of the bus. So if something happens and Jim slams on the brakes, it's your feet that take the hurt.
2. The farther back in the bus, the higher you bounce with each bump.
3. It's very womblike when the bus is rumbling down the highway—the constant throaty vibration lulls you beautifully.
4. But when the bus stops, and the engine turns off, you wake right up. The new city is disorienting—you went to sleep in Omaha and now you're in Denver. But it's okay. The bus is the bus is the bus, all across America.

My favorite morning was the last, I think. We were headed into Salt Lake City from Denver, and I woke up somewhere in Wyoming, under a sky that was Chamber of Commerce blue and with a skin of snow on the high desert. This was almost the first snow we'd seen. It had been preternaturally warm everywhere—when we'd left Boulder the night before they were busy evacuating 583 homes about fifty miles west, because of a roaring forest fire. In December. We plunged down into the great gap toward the shimmering valley of the Great Salt Lake, and it was not hard to imagine Brigham Young declaring, "This is the Place."

My first stop when we hit town was a bookstore. And herein lies a tale. When I'd last seen Tim DeChristopher, he'd been in federal prison in the high desert of California. A few weeks before our arrival in Utah, though, having served eighteen months of his sentence, he'd been returned to Salt Lake and released to a halfway house. Tim was allowed to leave for a job during the day, the better to ease his return to society. His church had offered him work helping with their ministry to poor people, but the Bureau of Prisons official had exploded at the news, bellowing at Tim that his was a "social justice" crime so he sure as hell wouldn't be "doing social justice work." As a replacement, then, Tim found work at Ken Sanders's antiquarian bookstore—which happens to be the only bookstore on the planet mostly devoted to the memory of the great desert writer Edward Abbey. It was piled high with copies of Abbey's most famous work, *The Monkey Wrench Gang*, a rollicking account of eco-sabotage that, in the 1980s, had given birth to Earth First!, the monkey-wrenching environmentalists who spiked trees and poured sand down bulldozer gas tanks. And on that very day we were in Salt Lake City, their brave spiritual descendants in Texas had climbed deep down inside the newly laid southern length of the Keystone XL pipeline and chained themselves to concrete blocks. Anyway, I knew Abbey a little, and admired him a lot, and I am reasonably sure that he would have roared with laughter to know that Tim, the most famous eco-criminal of his day, had been effectively sentenced to selling T-shirts with crossed monkey wrenches by a federal official equal parts vengeful and obtuse—a character straight out of Ed's novels.

Tim looked fit and calm and happy, and we talked about his plans to head east for divinity school come fall if his probation officer gave the okay. And then I spent the afternoon getting ready for the last of our shows. By this time I hardly needed to glance at the script—I'd gotten the hour and a half down pretty pat, knew the cues to the video clips and the laugh lines. In fact, I almost knew it too well—I'd found myself, the last few nights, starting to play myself playing myself, a kind of Hal Holbrook does Bill McKibben thing that unsettled me. It's hard to get truly psyched up to say the same thing night after night; on the other hand, this wasn't a play. I kept saying, "The nights on this tour feel like some of the most important nights of my life," and I wanted to really mean it.

Tonight would be no problem, and not just because it was the last. I'd wanted to come to Salt Lake in part because I was pretty sure Terry Tempest Williams would show up to help. She's not just among my favorite writers—her book *Refuge* is on the short list of the classic pieces of American nature writing—she's among my very favorite people. We'd known each other for a quarter century, since she had invited me out to Utah to give a talk when *The End of Nature* first appeared; her absolutely unfailing kindness, matched by that of her husband, Brook, is one of the givens in my life. So it didn't surprise me in the least when she volunteered, despite ill health, to make the long drive from Moab up to the university auditorium where we were gathering, and to give the sweetest, strongest possible introduction—a promise, in effect, that this crowd of her people would be part of this fight, though they'd be doing it in the most conservative state in the union. And in return I gave the truest talk of the whole long month, a valedictory that gathered in strength as it rolled toward its end.

I realized, as I was talking, how much the trip across America had touched me. It had been years since I'd driven the country; now, pressed for time, I usually flew, with America reduced to a two-dimensional map out the airplane window. So it had been a revelation again to sense the size and relief of the country: to start out amid those great volcanoes of the Northwest, to meander south through the coastal forest and toward the Hollywood hills and the Pacific beach—iconic for good reason; to drift through my beloved Northeast, from the last lingering leaves of the New England fall to the spreading concrete of the New York corridor, to the gentle Georgia hills; to strike out across the Appalachians, and into the rippling flat of Ohio and the dead flat of Illinois and Nebraska, and to come roaring across those plains and then run into the heart-filling wall of mountain west of Denver. We'd seen damage everywhere we'd gone—the ocean off Puget Sound too acidic for oysters, the ruined beach towns of the Jersey Shore, the sere droughty farmland of the West, even the smoke from those bizarre lingering Colorado fires. But it was still so beautiful, still so worth saving from the radical simplifiers of the fossil fuel industry who were crashing a million years of evolved gorgeousness and meaning into a homogenized layer of hot, bare, broken planet.

Everyone climbed up on the stage for the end—Terry, the great writer Rick Bass, who'd driven in from Montana, the young people who'd worked with Tim at his local group Peaceful Uprising, and our crew: Jean, who'd been bus mother all week, though she was the youngest on board; Rae, a sharp sly organizer; Steve, our silent, solid photographer; Duncan, a whiz both at organizing and at computers; and Matt, smiling Zen master of the whole operation. We, in turn, managed to get the whole crowd on its feet, ready to fight, ready to march.

We'd accomplished our objective. I'd written that piece for *Rolling Stone* in order to launch the tour, and the tour had launched a movement—we put out a press release in mid-afternoon announcing that we were well past one hundred campuses with active divestment fights under way. It wouldn't be easy at all—I knew the *New York Times* was planning an article for the next day, which would doubtless highlight the pitfalls and difficulties. But that's the nature of fights. We were, at least, in one; that much I could say for sure as I came off the stage. Also that I wanted a beer, and that I was ready to head for home.

If I thought the end of the tour would mean a rest, I was wrong—I'd been back in the house about an hour, unpacking the suitcase and drinking a beer, when the *Times* article popped up online. By the next morning it was on the front page of the business section, and before long it was the most e-mailed story of the day, and then of the week. This was fine by us, since it explained in ringing terms just how fast and far this new divestment movement was spreading. The students were at the "vanguard of a new national movement," the *Times* declared, drawing the parallel to the anti-apartheid fight of three decades earlier.

> Students said they were well aware that the South Africa campaign succeeded only after on-campus actions like hunger strikes, sit-ins and the seizure of buildings. Some of them are already having talks with their parents about how far to go.
>
> "When it comes down to it, the members of the board are not the ones who are inheriting the climate problem," said

> Sachie Hopkins-Hayakawa, a Swarthmore senior from Portland, Ore. "We are."

Two days later *Time* magazine's Bryan Walsh chimed in, ending his account like this:

> University presidents who don't fall in line should get used to hearing protests outside their offices. Just like their forerunners in the apartheid battles of the 1980s, these climate activists won't stop until they win.

The journalists were right to feel the momentum—by week's end, which also marked the close of the semester at most schools, we had divestment fights up and running on 252 campuses. Or rather, students had them up and running—at best we were helping to coordinate, to spread the word. "More vector than virus," was how I'd taken to describing my role. Still, it was a heady moment, for this was one of the biggest student movements in many years. And I was proud of my own college, Middlebury. Its president, Ron Liebowitz, in the midst of all the national furor, sent out a message to the whole campus promising a serious discussion in the semester to come. "A look at divestment," he wrote, "must include the consequences, both pro and con, of such a direction, including how likely it will be to achieve the hoped-for results and what the implications might be for the College, for faculty, staff, and individual students." Which seemed just about right, and set us up for a good fight on home turf.

In the middle of all the op-ed writing and tweeting and so forth, though, a phone call came from the most important address in all of academe: my daughter's freshman dorm room at Brown. And it wasn't a happy call—Sophie was reporting that, hours before her last final exams, an ambulance was en route to take her to the hospital because she'd been throwing up for hours. I jumped in the car and made the long drive to Providence. By the time I got there she'd been released and was ready to head home. A few days later, on New Year's Eve, surgeons removed a tumor from her belly. One cares about saving the world, but one *cares* about one's daughter; it was a long day in the hospital, exactly none of it spent worrying about Shell or Obama or Keystone.

For the next few weeks it was hard to get my head back in the game; once the holidays had passed and normal life resumed and the e-mail in-box was filling with its typical four hundred missives a day, I found myself shaky. Unnerved by it all. Overwhelmed. Frustrated and a little resentful. I was receiving a great deal of attention, and much of it was kind: the state's biggest paper named me "Vermonter of the Year," which was an honor that meant something to me. But my life didn't feel like my life. It was something I'd been sensing for the past couple of years, roughly the span covered by this book; I felt like the marble in the pinball machine, bouncing off one flipper after another. Or maybe I was the one playing the game, shaking the machine. I was good at this, after all; we were on about our fourth extra ball with Keystone. But it wasn't me, or at least it wasn't the me that used to be, the one that wrote difficult books, that had the time to figure things out instead of just reacting.

My reactions, as I say, were pretty good. There were plenty of moments when I felt as if I knew where to plant my foot next—I'd known how to scare the White House for a season, and the game plan for divestment was on schedule. The right words came at the right time. Maybe a lifetime of thinking was paying off in a few years of action. But some part of me was desperately thirsty for that older way of being. I felt unanchored, tossed on the tides of the heaving Internet. I understand that I could have lashed myself to the mast—either that or stopped up my ears; there's even software ("Freedom") that lets you block the Web from your computer. But successful activism seems to demand immersion in the moment, with small battles at every turn. It's hard to turn off the news cycle, at least for me; engagement is engagement. I'd never fixated on weekends before, because for a writer the "workday" is a fluid concept—but now the weekend meant that blissful period when e-mail dropped off to an almost manageable level. Real politicians, I think, thrive on this kind of stimulus, but I'm not a politician. A writer, if you think about it, is someone who has decided their nature requires them to hole up in a room and type. You can violate your nature for a while, but eventually it takes a toll.

Which is why I kept finding excuses to hang out at the farm with Kirk. He was in winter hiatus, mostly just resting, gathering strength. There'd been some more bear attacks on some of the distant beeyards ("I'm just going to have to electric-fence them all"), and a wild windstorm had blown over some of the colonies next to his house. But basically he was calmly reading, touring the woodlot on his old cross-country skis, and looking through seed catalogs. I've said before that he lives outside the world of the Internet, the TV, the cell phone—that's doubtless the oddest thing about him, statistically. He's a solid human being, attractively and somewhat dauntingly solid.

Months before he'd sent me home with a book to read, a

well-thumbed copy of *The Still Forest Pool* by the Thai Buddhist monk Achaan Chah, one of the people who'd brought insight meditation to the West. Kirk had bookmarked a small section titled "The Spiral of Virtue, Concentration, and Wisdom." "Virtue" and "wisdom" seemed unlikely, but "concentration" sounded seductive. And so one day after the new year, with Sophie recuperating nicely, I went by the farm for a ski and a long lunch of roast beef from one of the neighbor's cows. And I asked Kirk, quite directly, about his spiritual life.

"What you need to know is that I was very sick some years ago," he said. "As it turned out, I had mercury poisoning, but I didn't know that at the time. All I knew was that I couldn't sleep. I mean, one or two hours out of twenty-four. And I can't tell you how hard that is, how gray and horrible the world gets when you're that deprived of sleep. I called up one other guy who I'd heard had serious sleep trouble, and he sent me a book by Joseph Goldstein called *The Experience of Insight*. I think I'd heard about meditation before, I think I had a little book about yoga. But this was new—it was clear to me that it was not a way to make yourself feel better, but a way to know things about yourself. I recall reading in Epstein's book, or maybe Achaan Chah's book, the advice that you should 'try to stay awake longer and longer so you could meditate more.' And I remember thinking to myself, 'I've got that part already—I can't go to sleep.' That little piece of humor coming into my mind was important. You can't imagine how terrible sleep deprivation is, and when I was able to make that little teeny joke to myself, I think I kind of started to turn the corner."

Kirk learned to meditate. "I even went to a meditation retreat once," he said. "But one of my insights is that the way I live is a little like being on a retreat all the time." Indeed, he doesn't do much formal sitting anymore. "When I was, the meditation was like an island in me. It occupied a niche. But eventually it grew, it spread to the other parts of my life. And it felt like those other parts of my life became a more powerful form of meditation."

This set me up to ask the question that had been in the back of my mind all year: "But don't you get bored?" I mean, he spends his whole life dealing with his apiary. He lives in one large room; there are perhaps two hundred books along the wall. It's the Spartan opposite of the life I'd been leading these past months.

"I never get bored," he said after a moment's thought. "Sometimes I don't feel great, but I never get bored. The purpose of meditation, I think, is to be able to see the incredible beauty of life in every little aspect of it—so boredom is not my problem. My problem is not being overwhelmed by the amazing diversity of the world. I mean, one little piece of wood here on the wall of the house. I can remember the board it came from, and I think of my friend's sawmill, where I've bought all the wood for my colonies for years. The couple that runs it are in their seventies, but each winter they're out in their woodlot cutting the trees they'll need for the year. So I'm thinking of them, and of the forest. It's about stories. The real challenge of doing something like farming is to string all these stories together so they end up making sense. Farming is like playing a continuous game of 3-D chess. There are an innumerable number of moves, so many things are going on at so many different levels. A great farmer is one who can see what the right moves are and make them at the right time. It's completely absorbing. I try to figure out, every single day, what is the optimum purpose of this day? What is its best use?

"Take a single day from the year, just pick one from the calendar," he said.

"May fifteen," I said, for no good reason.

"Good choice," he said. "Spring has come, in fact it might be summer. Any day during May if I don't go to the beeyards I probably lose two thousand dollars in potential income. So much of the way the apiary produces honey depends on May—the honey-producing colonies, that's right when their population is growing the fastest, and when they're most likely to swarm. So it's really important to shepherd that population growth so you'll have a big colony to produce honey with. And it's right at the moment when customers are likely to be calling. But at the same time, it's a really key stretch for planting a garden crop or doing the first weeding. If you hoe a row of vegetables at the right moment, it will take a tiny fraction of the time it would to weed them two weeks later. I don't like to work six days on and take one off; I like to work when the weather is conducive. There are many times in the spring and fall when if you don't work on Saturday and Sunday it can take weeks to catch up. And so you have to weigh everything. The point is, it's not my schedule. It's the schedule of living things, what they need."

This isn't "farming" in the modern sense, of course; it's farming in an older way of looking at things. "I like the way the Japanese farmer and writer Masanobu Fukuoka put it," said Kirk. "'Farming is the cultivation of better human beings.' In another sense, there's really only one measure of good farming, and that's to leave the land better than you found it. If what you're doing is leaving the land more fertile, that's a pretty good guide for being a better human being. But it's so completely foreign to our cultural idea of using up resources, of mining the world and moving on to the next place, the next thing, of moving on to find some more."

It's this culture that Kirk, at some early point, decided to keep at bay—the one that seemed too confusing, too out of control. The very same culture that, in 2012, had managed to create the hottest year America had ever seen. The one that depends on more, on faster, on ambition, on a kind of generalized horniness for accumulation and sensation and novelty, novelty, novelty—novelty being the stock in trade especially of the Internet, the idea that at every second something new might be filtering into your in-box or onto Twitter, that something new might be showing up on YouTube or attaching itself to your Facebook timeline. And so he's kept it out, and concentrated on what was important.

And that calm has come at a cost in companionship. A high cost. "With most people, I feel like they'd bring in some of that craziness from the outside world," he said. "I mean, I've made a choice not to pursue material things, or recognition, or to try to fit in because everyone else was fitting in. And those are difficult things for people to do. I think most people in society are oriented around those things even if they don't want to be. And I've got no desire to impose my sense on others." It's not that there's no one he might have settled down with; doubtless there are some women who might have made his home a less lonely place without inviting in the whole culture. But maybe not so many—the odds weren't great, and he hadn't, so far, lucked out. It's not like he was going to go on eHarmony, after all. And the need had grown less urgent. "I used to say I was no more cut out to live by myself than anyone else, but I've definitely gotten better at it."

His house is, in fact, a slightly monkish cell, in the best sense of that word—the monk who looks out, engaged with but not overwhelmed by the world, the monk who stands solidly on his two feet, hard to shake. The one who comes from a tradition that he is carrying on. "I miss the old people so badly, the ones who lived before things like electronics were affecting our minds so much," he said. "They accomplished so much more on a personal basis, and they thought nothing of it. Think of all the small farmers in the late nineteenth century who figured out modern beekeeping, all of the things we still do." His order is the order of farmers, true farmers, the people who stay at home and make home better. "When I was in the middle of those health problems, when I couldn't sleep—I really thought the little flame of my life was about to go out. I just asked, if I could recover, I said I'd devote the rest of my life to helping restore the world of nature. That was my sincere desire from that place, and I really have felt since that I've had to keep that promise."

As it turned out, of course, the cause he wanted to devote himself to was not just what he needed to make a meaningful life, it was what the planet requires if we're going to make it through against greater and greater odds. It's what, I suppose, I've devoted my own life to, as different as that life has been in recent years. My constant motion and his fixed gravity grow from something similar. Or I hope they do. "For us to survive now we really have to put other living things ahead of ourselves," he said, as we ate the last of the beef and the last of the cabbage slaw, and drank the last of the beer. "That goes back to meditation, to the idea that we're not *really* a self, that we're connected to the world in so many ways. Since everything's connected, maybe that's why I can live on my own. Because my life makes sense to me, because it adds up."

As for me, the sense that I was living an unnatural life deepened as the winter wore on. I'm not prone to depression, but I found myself often blue, in part, I think, because home was not the refuge I remembered. I spent more time than usual in Vermont in the winter of 2013, and I'd been counting on it to restore me. But fights seemed to follow me back home.

I'd helped launch divestment campaigns across the country, but the one that mattered to me was at Middlebury, the school where I'd taught for a decade, and whose ways I'd come to love. It's on the short list of truly great small liberal arts colleges in the country: the kids are relentlessly smart and highly engaged. And because it's perched between the Green Mountains and the broad farm valley that stretches to Lake Champlain, it attracts students with a bent toward the outdoors. It's no accident, I think, that it has the oldest environmental studies department in the nation, and no surprise that its students quickly mounted a powerful divestment campaign, convincing two-thirds of the student body to back divestment. The president, who had led the college ably for years, moved quickly to engage the topic, inviting me to speak on a panel discussion shortly after Christmas break.

I trusted him completely, and I was impressed by the discipline and hard work of the kids leading the divestment charge. But I also knew that Middlebury, like many institutions, had a board dominated by professional investors; since that's where the money is now, Wall Streeters tend to dominate most high-powered boards. (Middlebury has had rough luck, in fact, with some of its high-profile trustees: Dennis Kozlowski, the CEO of Tyco, for instance, went from the Middlebury board to a New York State prison, after he had his company help pay for a fortieth-birthday party for his second wife where an ice sculpture of Michelangelo's David peed vodka for happy guests.) The trustees I knew personally were terrific—not just generous, but remarkably engaged in trying to push the college forward. They mentored students, designed green buildings, and played a larger role in campus life than at most schools. But I also knew they'd take "fiduciary duty" with the Gospel seriousness of lifetime investors.

So I prepped hard for the panel and didn't spend overmuch time harping on the dangers of global warming. I spoke last, after four investment professionals, and so we may have been behind on points—but I had a couple of aces in my sleeve. One was a brand-new report from a research firm, Aperio Consulting, that had done a thorough study of just how much divestment would have cost the average portfolio over the past decades. The "theoretical return penalty" of excluding fossil fuel stocks, they concluded, was 0.0034 percent, or about as close to zero as one could get.

Better yet, I had a letter from a guy I'd met the summer before. His name was Tom Steyer, and he'd called me out of the blue after reading the *Rolling Stone* piece, introducing himself as an investor deeply concerned about climate change. He insisted on flying across the country from his San Francisco home to talk about it. He'd mentioned he was a hiker, so I agreed to go for a climb with him—if he was a rich bore, at least we'd be up a mountain and the day wouldn't be entirely wasted. As it turned out, he could hike as fast as I could, and he was pretty interesting. He'd founded Farallon Capital Management, one of the country's richest hedge funds, in the process making himself enough money to qualify for the Forbes 400 list. But somehow he'd remained a progressive, and a passionate one. His wife, Kat Taylor, who I'd meet later, ran a community development bank in Oakland; they'd given small fortunes to Stanford and Harvard and Yale; he'd paid for and run several ballot initiatives in California to help push green causes; he'd just given a speech at the Democratic National Convention; he was being talked about as a possible secretary of energy in the second Obama administration. He had, that is, something more important than money for our fight, which was credibility among the financial elite. And he was willing to put it to use.

And one way he put it to use was by writing a letter to the Middlebury trustees. I couldn't resist introducing it with a bit of a flourish. "This is someone who's a better investor than anyone in this room," I said. "How do I know that? He's made more money." Indeed, his net worth was half again as large as the college's $900 million endowment. And Steyer explained not only that divestment made moral sense, and not only that it would enhance the college's brand and image, but also something else that hadn't really occurred to me. "I believe a fossil fuel free portfolio is a good investment strategy," he wrote. "The data on climate change makes it clear something has changed, and as the rest of the world realizes this, fossil fuel stocks will come under increasing pressure. At the moment, other investors have not fully realized the risk that carbon reserves will become a stranded asset; if you acknowledge what your own science departments are telling you, this give you an edge relative to those investors. I can tell you that in my own investments, I have directed my financial team to divest my holdings of fossil fuel investments—in part because I am convinced it will outperform the market."

Playing that card felt good—in fact, I had my friend Jon Isham, an economics professor who'd worked for years on climate questions, hand out copies of the letter to the whole overflow crowd. People went away understanding that the college could, in fact, divest without facing ruin. But, as successful as it was, the evening was a reminder that we were asking people across the country to do something difficult—making demands on the people who pay your salary or your scholarship, and who you like and respect, is psychologically harder than making demands on your government.

Harder still was knowing that a small but vocal portion of my Vermont neighbors were really angry at me. If Vermont is going to produce renewable energy, some of the power will need to come from wind; we're far enough north that solar panels alone (though I have them across my roof) won't provide year-round power. And on this edge of the continent, the wind blows hardest at higher altitudes, which is why the country's first commercial wind turbine went up on a Vermont ridgeline during World War II. But recent attempts to put wind farms on a few of the state's ridges have met with vociferous opposition—the towns with the turbines have supported them, in part because of the tax payments, but surrounding neighbors have complained loudly: about the sight, about damage to birds, about effects on human health from the sound of windmill blades. That winter they were mounting an effort to put a three-year moratorium on new windmill construction—the installations, they said, were simply out of scale with Vermont's landscape.

Since I'd been named Vermonter of the Year on New Year's Day, the Speaker of the House summoned me to the legislature in mid-January to address the whole body, a rare honor. And having the podium, I said what I felt about the moratorium plan: it was well-intentioned, and it was a mistake. We were desperately fighting the Republicans in Congress to maintain any kind of funding for renewable energy; if Vermont declared it was too precious for windmills, that fight would get impossibly hard. Mostly, of course, I talked about climate change in general, and the impact of Irene, and the need to weatherize homes. But the paragraph on wind turbines was what made the news the next day:

> I recall, last year, when the leading opponents of "big wind" in Vermont told reporters that "they are making climate-change victims out of the people who live around the projects," and that it was akin to "burning villages in order to save them." Let me say that I think such statements are incorrect. Climate change victims are, say, the 150,000 displaced from their homes on Monday in chaotic flooding in Mozambique. We know the devastation that came with Irene—imagine a quarter of the population of Vermont displaced. Burning villages can actually be found, in places like Tasmania or Colorado, where record wildfires in the last year have taken lives and wrecked communities. So I think we should plan carefully but quickly to minimize the ecological footprint and maximize the energy gain. That energy gain is real: every spin of that windmill blade reduces the need,

> somewhere, for burning coal or gas or oil; in New England, first of all, where we still have lots of fossil-powered electricity being generated. But it also reduces by some small amount the pressure on a Bangladeshi peasant farmer or a doctor fighting the spread of dengue fever. We do not need them, as I say, on every ridgeline, but I continue to hope for the day when I see them on top of Middlebury Gap, the ridgeline above my home, turning with slow and stately beauty, the breeze made visible and the future illuminated.

It didn't surprise me when I got a slew of nasty e-mails—as nasty, some of them, as I was used to getting from right-wing climate deniers. But it did shock me a little when I went to the library one town over to give a little book talk and found the mother of one of my daughter's former junior high school classmates distributing flyers accusing me of having too large a carbon footprint because I travel around the world organizing 350.org. I knew most of my neighbors didn't feel the same way—in a poll, I'd have done okay. But polls are for politicians, who don't seem to mind having 49 percent of people angry at them if 51 percent feel the other way. Most normal people, me included, don't really enjoy controversy.

I'd slowly sucked myself in, going from writer to global educator to unlikely and somewhat reluctant activist. At every step the controversies multiplied. No complaints—if you're going to hit, you're going to get hit back. But I didn't thrive on the combat—it made me a little sick to my stomach. Adulation was a little easier, but not a lot. Real politicians, I've noticed, love to work a crowd, drawing energy from everyone they met. I'm a writer; left to my own devices, I'll retreat to my room and type. I wasn't cut out to be a leader.

Still, I was one. And since one of the things leaders do is rally people, it was time to head back to Washington, where this story began. We'd been planning, with our friends at the Sierra Club, for a big D.C. action, and we knew that with a Keystone decision looming we'd have to somehow organize in midwinter. There's a reason that big Washington rallies come in the spring, summer, and fall. And for a global warming rally there's also the question of optics; nothing like a good snowstorm to cover up the message. But it was a risk we had to take.

In some ways, simply involving the Sierra Club was the biggest victory. From the start at 350.org, we'd said we weren't trying to create a big organization—we wanted to build a movement, create campaigns where everyone could play, and break down the walls between groups, and between insiders and the grass roots. That's easier said than done—the Sierra Club and other large outfits come with their own internal politics and histories; they need to keep a separate profile to raise money; every group is proud of its own distinctiveness. But from the first day of the Keystone arrests, the key big environmental groups had started coming slowly together in a new way. And Michael Brune of the Sierra Club was moving particularly fast. Among other things, he'd been urging the group to drop its 120-year prohibition on engaging in civil disobedience, and early in January he persuaded the board to grant a trial run. So we scheduled a week of climate action for mid-February—a small and controlled day of arrests outside the White House, and then a massive (we hoped) march for President's Day weekend.

Our friends at the Hip Hop Caucus joined in the call. Their leader, Rev. Lennox Yearwood, had been a huge help on the Do the Math tour. He was the opposite of the traditional environmentalist. (Forget John Denver; think Frank Ocean.) He and his crew brought a special focus on people of color, 350.org provided youthful energy, and the Sierra Club gave us the link all the way back to John Muir. And so we went to work, doing the grinding task of making sure there were buses and permits and sound systems and porta-potties.

One of my jobs was helping round up prominent people to go to jail. The U.S. Park Police were clear: if we came with more than fifty, there'd be no arrests. We knew we wanted to highlight some of the folks who'd been blockading the Texas segment of the pipeline, and some Nebraska ranchers, and some survivors of Sandy, and some leaders of the spreading divestment movement, and that left us about twenty-five spots to fill. It was enlightening to watch some prominent people squirm when we asked; they were eager to do something—just not something that would get them in trouble with the administration, whose representatives were working the phones themselves, trying to persuade us to call off the whole affair. But when we gathered that morning in Lafayette Square, the park where I'd spent those two tumultuous weeks in August 2011, we had a pretty fine crew. There was a billionaire, and there was a Kennedy (two, actually—Bobby Jr. brought his son Conor, who because he had dated the pop star Taylor Swift got us all kinds of coverage in unlikely outlets). There were preachers and environmental leaders, a former poet laureate and a solar pioneer and a prize-winning climate scientist. For many of us, though, the biggest thrill came when we were joined by Julian Bond, the former head of the NAACP. I was handcuffed across from him in the paddy wagon, and got to listen as he told of the days in 1960 when as a young college student he went to jail for helping to desegregate the lunch counters of Atlanta. You could feel the movement broadening, deepening, joining the current of change that has always run through the country when it needed it most.

And you could feel it even more three days later, when we rallied on the National Mall. The day dawned cold—well below freezing, and with a twenty-mile-an-hour wind that I knew would sap people standing in the open. We weren't going to get any casual passersby to this event—it would be a real test of what kind of movement we'd built.

By noon, when I got up on stage to open the rally, I knew we'd passed. Buses had been rolling in all morning, from thirty states stretching all the way to Minnesota (in cities across the West, meanwhile, big solidarity rallies were getting under way). The Mall was filled as far as the Washington Monument in the distance, a sea of people far larger than had ever assembled for a climate rally before. The U.S. Park Police don't actually estimate crowds, but I overheard one sergeant arguing with our police liaison that we'd exceeded our 50,000-person permit; the press, which gave the rally wall-to-wall coverage, settled on "at least 40,000" as the best guess.

The speakers were admirably brief in the cold, and fierce: Van Jones, who had worked in a high-profile White House job, told his old boss that Keystone was the most important decision he'd ever make, and that if he approved it, "the first thing that pipeline will run over is the credibility of the president of the United States"; Tom Steyer explained that he was an investor and that Keystone was a bad investment; a delegation of Alberta native leaders led by Crystal Lameman, who, after all, had fought this fight longest of all, explained once more the stakes for their people and for the planet. Senators and celebrities, singers and scientists—everyone was swaying happily in the cold as Reverend Yearwood rocked the mic, reminding us that this was "our Birmingham, our lunch counter moment."

When it was my turn to talk, I did my best to focus on Keystone, too—that was the issue at hand, the fight I'd chosen. But looking out over the vast crowd, the largest I'd ever addressed, I knew it represented much more than anger at a single pipeline. I knew the size of the crowd meant that people in large numbers had finally managed to overcome the numbing sense that there was nothing to do about global warming. "All I ever wanted to see was a movement of people to stop climate change, and now I've seen it," I said as I began to talk. "You are the antibodies kicking in, as the planet tries to fight its fever."

As I talked, I thought of the young man many years before who'd first read and written about climate change—a young man who was now much older. I thought of the young people I'd started 350.org with—they were older, too, and I could see them scattered around the stage smiling up at me, and that made it easier to speak in front of this vast throng. And I thought of Kirk, doing his work patiently back home, and I thought even of the bees, doing *their* work for countless millennia. There'd been stories in the paper the past few days about how 2012's crazy weather had wrecked the honey crop around the world—record rainfalls in the United Kingdom had cut British harvests by 72 percent, while drought had produced the worst results ever in Australia, New Zealand, and Spain. There were no flowers, and hence no nectar. Too much oil, too little honey. We'd waited so long to get started with this fight—maybe too long.

But we *were* started, and looking out on that field I realized one reason, maybe the main reason, that I'd worked so hard. I didn't particularly want to lead a movement, but I wanted to join one, and so I'd helped to build it.

"I can't promise you we're going to win," I finished. "But I've waited a quarter century, since I wrote the first book about all of this, to see if we were going to fight. And now, today, at the biggest climate rally in U.S. history, I know we will. The battle—the most fateful battle in human history—is finally joined. And we will fight it together."

I got home the next morning, flying in over those mountains that have bordered my life—the Dixes, and Giant with her long steep slides, and Mount Mansfield hovering to the east. The ground in the valley was white on that February morning, and spring was still a ways away, at least in the old order of things. I knew the hives I flew above were full of resting bees, and I knew they'd soon be in flight. The old cycle we've always known is very nearly gone, but not quite. It lingers yet, and while it does the fight is worth the cost.

AFTERWORD

In the months since the hardcover publication of this book, not enough has changed in the world: a decision over the Keystone XL pipeline remains delayed, and serious action on carbon emissions has been postponed for another year in almost every nation.

But one thing that's shifted, a little, is my thinking on at least one facet of this struggle. As the pages of this book make clear, I've spent the last few years in constant motion around the country and the earth. I'd come to think of myself as a "leader"; much of this book reflects on that growing identity.

In recent months I've come to like that idea of leaders less and less—it seems to me to miss the particular promise of this moment, which is that we could conceive of and pursue movements in new ways. In particular, for environmentalists, we have an analogy close to hand. We're struggling to replace a brittle and top-heavy energy system, where a few huge power plants provide our electricity, with a dispersed and lightweight grid, where ten million solar arrays on ten million rooftops are linked together. The engineers call this "distributed generation," and it comes with a myriad of benefits. It's not as prone to catastrophic failure, for one. And it can make use of dispersed energy, instead of relying on a few pools of concentrated fuel. The same arguments seem to me to apply to movements.

My wife Sue Halpern and I spent much of the summer of 2013, for instance, crisscrossing the country for a series of rallies called Summerheat, that 350.org was helping support. We didn't organize them—we knew great environmental justice groups all over the country, and we knew we could highlight their work, and highlight the links between, say, standing up to the toxic Chevron refinery in Richmond, California, and standing up to the challenge of climate change. From the shores of Lake Huron and Lake Michigan where there's a proposed tar sands pipeline, to the banks of the Columbia River where they want to build a big oil port in Vancouver, Washington, from Utah's Colorado Plateau where they've proposed America's first tar sands mine, to Sebago Lake in Maine where they want to reverse the flow of a pipeline to carry tar sands, from the coal-fired power plant at Brayton Point on the Massachusetts coast to the fracking wells of rural Ohio: Summerheat demonstrated the local depth and global reach of this emerging fossil fuel resistance. I had the pleasure of going to talk at all these places and more besides, but I wasn't crucial to any of them—a pollinator, not a queen bee.

Or consider the Keystone. In 2012 the *Boston Globe Magazine* put a picture of me on the cover under the headline: "The Man Who Crushed the Keystone Pipeline." I've got an all-too-healthy ego, but even I knew that was over the top. I'd played a role in the fight, writing the letter that asked people to come to Washington to fight the pipeline, but it was effective because I'd gotten a dozen friends to sign it with me. And I'd been one of 1,253 people who went to jail, in what was the largest civil disobedience action in this country about anything for years—it was their combined witness that got the ball rolling. And once it was rolling, the Keystone campaign became the exact model of the distributed, loosely linked power network I'm describing.

The big environmental groups played key roles, supplying lots of data and information and keeping track of straying Congresspeople: NRDC and Friends of the Earth, the League of Conservation Voters and the National Wildlife Federation, on and on, none of them spending time looking for credit, all of them just pitching in. But the groups on the ground were equally crucial: the indigenous groups in Alberta and elsewhere that had started the fight and graciously welcomed the rest of us without complaining about how late we were; the ranchers and farmers of Nebraska, who roused the whole stadium at a Cornhuskers game to boo a pipeline commercial; the scientists who wrote letters, the religious leaders who conducted prayer vigils. Bloggers appeared; one upstart Web site won the Pulitzer Prize for its coverage of the tar sands.

Other people became specialists in State Department dealings or in the chemical composition of bitumen; CREDO (half activist organization, half cell phone company) has now signed up 75,000 people pledged to civil disobedience if the pipeline gets approved. There's the Hip Hop Caucus, whose head Lennox Yearwood has roused one big crowd after another, and there are labor unions—nurses and transit workers, for instance—who've had the courage to stand up to the pipeline workers union that would benefit from the small number of jobs that might be created were Keystone built. There are groups of Kids Against KXL, and there was a grandparents march from Camp David to the White House. Some of the most effective fighting has come from groups like Rising Tide and the Tar Sands Blockade in Texas, organizing epic tree-sits to slow construction of the southern portion—but the Indigenous Environmental Network has been every bit as effective in demonstrating to banks the folly of investing in the tar sands. First Nations people and British Columbians have blocked one of the proposed pipelines to the Pacific; inspired activists have kept the filthy oil out of the European Union.

Whatever the final outcome of the battle, it's clear that this kind of full-spectrum resistance can stand up to the huge bundles of cash that are the industry's only argument. And in the course of the campaigning, we change each other: it was inspiring to watch the Sierra Club, the nation's oldest and biggest environmental group, which for years had helped lead the fight against coal power, decide to change its 120-year-old bylaws to allow it to take part in civil disobedience.

In fact, this sprawling campaign exemplifies the only kind of movement that will ever be able to stand up to the power of the richest industry the planet has ever known. In fact, it will have to get much much larger still, incorporating its obvious allies in the human rights and social justice arenas. (There's never been a clearer threat to survival, or to justice, than the rapid rise in the planet's temperature caused by a small percentage of its citizens; conversely, we can't have a real answer to our climate woes without addressing the insane inequalities and concentrations of power that help drive this and so many other disasters.) That's why it's such good news when people like Naomi Klein or Desmond Tutu join in the climate struggle—it makes it clear that it's not, in the end, an environmental battle at all, but an all-encompassing fight about power, hunger, the future. And just as it needs to expand by category, it also needs to grow by geography. In 2013, 350.org and its allies trained 500 young people from 135 countries in Istanbul, and they are now organizing conferences and campaigns in their home countries. Only by expanding exponentially can we ever get big enough to succeed. And that means that the value of particular leaders is limited at best.

It doesn't mean, of course, that some people don't have more purchase than others in this movement. Sometimes that standing comes from living in the communities most affected by climate change or fossil fuel depredation. When the big climate rally finally did happen on the National Mall in April 2014, the 50,000 in attendance may have been most affected by the words of Crystal Lameman, a young member of the Beaver Lake Cree Nation, whose traditional territory has been poisoned by tar sands mining. Sometimes it comes from charisma: Van Jones may be the most articulate and engaging environmental advocate ever. Sometimes it comes from getting things right for a long time: Jim Hansen, the greatest climate scientist, gets respect from everyone I know, even if they disagree with him about, say, nuclear power. Sometimes it comes from sacrifice: Tim DeChristopher went to jail for two years. Sometimes it comes from dogged work on solutions: Wahleah Johns and Billy Parish figuring out how to build solar farms on Navajo land and crowdfund panels on community centers. Sometimes truly unlikely figures emerge, like Tom Steyer, whose influence continues to expand.

But it doesn't strike me that even figures like these are exactly "leaders" in the sense we've traditionally imagined—that is, those who chart the path for the movement. It's more, to use an analogy from the Internet age, as if they're well-regarded commenters. Or to use a slightly more universal image, they're elders. Not in a chronological sense necessarily, but in the sense that you turn to them for their take on things. Elders don't tell you what you must do, they tell you what they must say. And sometimes it is literally a matter of words. A few of these elders are, like me, writers; many of them they have a gift of condensing and crystallizing the complex. When Jim Hansen calls the Alberta tar sands the "biggest carbon bomb on the continent," it gets across.

Another privilege that comes with that standing is that you can get people to listen to your ideas. So when Naomi Klein and I hatched the plan for a fossil fuel divestment campaign in 2013, people paid attention (and all the more so when Archbishop Tutu lent his sonorous voice to the cause). People trust that you're not harebrained, not going to waste their energy on a pointless task.

And these elders-of-all-ages also play a sorting role, backing the ideas of others. There are days when I feel like the most useful work I've done is spread a few good Kickstarter proposals via Twitter, or written a blurb for a fine new book. And conversely, one can try to downplay ideas that strike one as weak. I was speaking in Washington in April 2014 to a group of grandparents who had just finished a seven-day climate march from Camp David. A young man demanded to know why I wasn't backing the sabotage of oil company equipment, which he thought was the only way we could ever damage the industry. I explained that I believed in nonviolent action, that we were doing real financial damage to the pipeline companies by slowing their construction schedules and inflating their carrying costs, and that in my estimation wrecking bulldozers would play into their hands.

But maybe he was right. I don't actually know, which is why it's a good thing I'm not the boss of the movement—no one is. Remember all those solar panels on all those roofs: it's distributed.

I'm sure much of this thinking is old news to people who've been building movements for years. I haven't—I found myself, or maybe stuck myself, at the front of one, and these thoughts reflect that experience. But I sense our job is to rally a movement big enough to stand up to all that money. It needs to stretch to every place with a thermometer; it needs to engage every fossil fuel company; it needs not just to prevent pipelines but to build windmills. It needs to remake the world in record time. That can't happen with a paramount leader, or even dozens of them—it can only happen with a spread-out and interconnected movement, a citizenry engaged. Rooftop by rooftop, we're aiming for a different world—one that runs on the renewable power that people produce themselves in their communities, in small but significant batches. The movement that will get us to that new world must run on that kind of power, too.

A NOTE FROM THE AUTHOR

The problem with writing a book of this sort is that, being a kind of memoir, it focuses on one person's experiences to the exclusion of so many who played as large a role or larger. I feel this particularly in the sections of this book that describe the Keystone Pipeline fight and the rise of the divestment movement. Those battles have become so broad, and are being fought so ably by so many people, that they would be better served by a real history, which I hope someone will someday write. I'll play a small part in those stories, but it's one I'll always look back on with (tired) fondness because of the remarkable people I met along the way.

ABOUT THE AUTHOR

BILL MCKIBBEN is the author of more than a dozen books, including *The End of Nature, Eaarth,* and *Deep Economy*. He is the founder of the environmental organization 350.org and was among the first to warn of the dangers of global warming. He is the Schumann Distinguished Scholar at Middlebury College, a fellow of the American Academy of Arts and Sciences, and the 2013 winner of the Gandhi Peace Award.

Books by Bill McKibben

Wandering Home—In one of his most personal books, Bill McKibben invites readers to join him on a hike from his current home in Vermont to his former home in the Adirondacks. Over the course of his journey McKibben meets with kindred spirits committed to the preservation of nature. For McKibben, there is no better place than these woods to work out a balance between the wild and the cultivated and to discover the answers to the challenges facing our planet.

Eaarth—Bill McKibben has a stark and sobering message: We've waited too long to stop the advance of global warming, and massive change is already underway. We've created a new planet; we may as well call it Eaarth. Our hope depends, McKibben argues, on building the kind of societies and economies that can hunker down, concentrate on essentials, and weather trouble on an unprecedented scale. Fundamental change is our best hope on a planet suddenly and violently out of balance.

Deep Economy—In this powerful and provocative manifesto, Bill McKibben offers the biggest challenge in a generation to the prevailing view of our economy. For the first time in human history, he observes, "more" is no longer synonymous with "better"—indeed, for many of us, they have become almost opposites. McKibben puts forward a new way to think about the things we buy, the food we eat, the energy we use, and the money that pays for it all.

Enough—McKibben turns his eye to an array of technologies that could change our relationship not with the rest of nature but with ourselves. He explores the frontiers of genetic engineering, robotics, and nanotechnology—all of which we are approaching with astonishing speed—and shows that each threatens to take us past a point of no return. This wise and eloquent book argues that we cannot forever grow in reach and power—that we must at last learn how to say, "Enough."